Animal Curation

The Science of Care and Welfare

Animal curation is a vital and evolving discipline that integrates science, policy, and hands-on care to ensure the highest standards of animal welfare. As the role of zoos, aquaria, sanctuaries, and research facilities expands beyond exhibition to conservation and education, the management of animals under human care has become increasingly scientific. This book provides a comprehensive guide to the organisation, policies, and procedures essential for effective animal care programmes. It emphasises evidence-based practices in husbandry, veterinary care, and facility management while prioritising both animal well-being and staff safety. Through detailed chapters and real-world case studies, readers will explore species-specific needs, ethical considerations, and regulatory compliance. Designed for students and professionals in animal science, welfare, and conservation, this book moves beyond basic care, focusing on the concept of 'thriving' rather than mere survival. It is an essential resource for shaping the future of animal management and welfare.

Andrew R. Halloran is an Animal Care & Welfare consultant dedicated to helping zoos and sanctuaries enhance the well-being of animals under human care. Halloran co-founded the Tonkolili Chimpanzee Project, a conservation initiative in Sierra Leone which aims to mitigate conflicts between humans and chimpanzees in anthropogenic landscapes. He is also the author of *The Song of the Ape* (St Martin's Press, 2012) and *Lion Shaped Mountain* (Elgin Press, 2021).

Animal Curation

The Science of Care and Welfare

ANDREW R. HALLORAN

Elgin Centre, Inc., Animal Care & Welfare, Vero Beach, Florida

CAMBRIDGE
UNIVERSITY PRESS

Shaftesbury Road, Cambridge CB2 8EA, United Kingdom

One Liberty Plaza, 20th Floor, New York, NY 10006, USA

477 Williamstown Road, Port Melbourne, VIC 3207, Australia

314–321, 3rd Floor, Plot 3, Splendor Forum, Jasola District Centre, New Delhi – 110025, India

103 Penang Road, #05–06/07, Visioncrest Commercial, Singapore 238467

Cambridge University Press is part of Cambridge University Press & Assessment, a department of the University of Cambridge.

We share the University's mission to contribute to society through the pursuit of education, learning and research at the highest international levels of excellence.

www.cambridge.org
Information on this title: www.cambridge.org/9781009181532

DOI: 10.1017/9781009181525

First published 2026

A catalogue record for this publication is available from the British Library

Library of Congress Cataloging-in-Publication Data
Names: Halloran, Andrew R. author
Title: Animal curation : the science of care and welfare / Andrew R. Halloran.
Description: Cambridge ; New York, NY : Cambridge University Press, 2026. |
Includes bibliographical references.
Identifiers: LCCN 2025018972 (print) | LCCN 2025018973 (ebook) |
ISBN 9781009181624 hardback | ISBN 9781009181525 ebook
Subjects: LCSH: Captive wild animals | Animal welfare
Classification: LCC SF408 .H35 2026 (print) | LCC SF408 (ebook)
LC record available at https://lccn.loc.gov/2025018972
LC ebook record available at https://lccn.loc.gov/2025018973

ISBN 978-1-009-18162-4 Hardback
ISBN 978-1-009-18153-2 Paperback

For EU product safety concerns, contact us at Calle de José Abascal, 56, 1°, 28003 Madrid, Spain, or email eugpsr@cambridge.org

IN MEMORIAM
Dr Douglas Broadfield
1966–2024
My advisor and friend. He was truly a lighthouse keeper, always able to
provide a guiding light even when the seas were rough.

Contents

Part IV   Operations

Conclusion

Introduction

Curatio Fundamentorum **1:** All species are unique, with unique needs, unique challenges, and unique measures of thriving.

Introduction

Figure I.1 The author playing his concertina for Allie. (Photo credit: author's collection).

I.1   The Ape and the Mouse

A frantic voice comes over the radio. The sound startles me. I am just a few months into my new job as the Director of Chimpanzee Behavior & Care at one of the largest chimpanzee sanctuaries in the world. I am struggling in my new position, unable to really figure out what my role is here. I am sitting in a meeting. Just as the fluorescent lights, white walls, and monotone conversation begin to lull me to sleep, I hear the radio. Secretly hoping that it's for me so I can leave this room, I sit up. The frantic voice asks for me to come to one of the chimpanzee buildings and bring 'something really good'. I excuse myself from the meeting and head to the kitchen, where I find a box of strawberries.

I arrive to find a chimpanzee named Allie (Figure I.1) behind the mesh wall of her enclosure, holding a small brown mouse. Across from Allie, on the other side of the mesh, an exacerbated care technician is looking at me and shrugging her shoulders. Allie is very gently cradling the mouse while stroking his fur. She looks at me, then down at the mouse. She slowly brings the mouse to her mouth and kisses him. The mouse, by the way, is clearly terrified.

Allie spent the first part of her life as a pet (Figure I.2). She was raised by a very loving human family who had her eating from a high chair, brushing her teeth, and being tucked into bed each night with bedtime prayers. However, as cute baby chimpanzees do, Allie grew into a very large and unpredictable adult. Despite being raised as a human, it became increasingly clear to the family that Allie was very much a chimpanzee, with chimpanzee strength, chimpanzee behaviours, chimpanzee mood swings, and chimpanzee needs. Because of this, Allie found herself being sent back to the chimpanzee breeding facility she had come from. There, she lived alone.

When Allie came to the sanctuary, she had very little experience with other chimpanzees. In fact, it is safe to assume that Allie didn't really even know that she was a chimpanzee. Since she had spent her life with humans, it is probable that Allie thought of herself as another human. This was evident the night she arrived.

I was there to greet her when she first came to the sanctuary. It was about 3:00 in the morning when she was brought into our special needs facility. Allie quickly came over to the mesh to meet me. I sat with her for the remainder of the night. As we watched 1970s Chicago videos on my phone, it was clear that Allie was more

Figure I.2 Allie with a purse. (Photo credit: author).

comfortable with me than the other chimpanzees across from her. There we sat, both of us new arrivals, thrown into a very foreign world, neither of us knowing what to expect next.

Now, months later, Allie has been moved to one of the buildings in order to be introduced to a group of other chimpanzees. However, like me, Allie has still not acclimated to this place. As Allie proceeds to create a little nest out of hay for her new pet mouse, I look over at the other chimpanzees who are acting like … well … chimpanzees. They are vocalising to each other, grooming, fighting, reconciling, and enjoying the ability to live according to their true nature. Meanwhile, Allie has placed the mouse in the bed she has made for her and covered her gently with a blanket.

As I watch this, I have an epiphany about Allie's care and about my own role at the sanctuary. Allie, as a chimpanzee, has very specific needs based on her species. Chimpanzees, by nature, are exploratory, self-aware, and extremely intelligent. In the wild, they are faced with complex choices throughout their day. They have been gifted with the ability to make decisions, not just by instinct, but by what they have learnt through social transmission with other chimpanzees. They survive by this ability. They are driven by their evolutionary need for survival to rely on socialisation, participate in learning by social transmission, and make decisions based on what they have learnt. Therefore, chimpanzees *need* a rich social environment. Chimpanzees *need* the ability for freedom of choice and self-determination. Chimpanzees *need* cognitive stimulation throughout their day.

However, Allie is also an individual with a unique individual history. This history has created severe challenges to her ability to have her needs met as a chimpanzee. Allie, who was raised as a human, and who has had no experience socialising with another chimpanzee, is not going to know how to function in a chimpanzee group. Allie, whose every movement has been either restricted or managed by a human, is going to have a difficult time acclimating to the freedom of choice that a large sanctuary habitat will grant her.

It will take a careful plan to grant Allie the life she deserves, and implicit in the definition of 'sanctuary' (Figure I.3). This plan will need to be created by those with an expertise in chimpanzee behaviour and animal husbandry. This plan will need to be executed by an experienced and educated staff. This plan will need to be consistent. Allie's welfare and progress will need to be continually monitored and assessed by both behavioural observation to ensure her psychological well-being and medical observation and examination to ensure her physiological well-being.

At the same time, the environment of this group of chimpanzees should be continually assessed and monitored, to ensure that it is meeting both the species-specific and individually specific needs of each chimpanzee. The environment should promote species-typical behaviour and movement. The environment should be cognitively stimulating. The environment should be a safe place for both the chimpanzees and the staff working within it.

As this epiphany dawns on me, I begin to realise my role within my new job. It is up to me, as the person in charge of the care of this chimpanzee, to create an

Figure I.3 Allie in a tree. (Photo credit: author).

environment for her where she can thrive, not just live. This environment must contain all of the variables that would allow her to live a social, cognitively stimulating, and self-deterministic life. These variables will be unique to Allie. I can only understand these variables by observing and assessing Allie's behaviour and indicators of welfare, while at the same time drawing on information from her veterinary examinations. This is in addition to the basic husbandry she requires – proper nutrition, a sanitary environment, and good veterinary care. The team I assemble to create and maintain this environment must operate safely. As the Director of Chimpanzee Behaviour & Care, I am mandated to both provide this environment, and ensure its continued functionality. This is not just true for Allie's care, but true for the care of every individual chimpanzee residing at the sanctuary. I smile at Allie, who has started to take an interest in the Pop-Tart in my left hand. She leaves the mouse and walks over to the mesh. The mouse seizes the opportunity and runs away. Allie enjoys her treat. She makes eye contact with another chimpanzee; a male named Spike, who is interested in what she is eating. He vocalises to her. She returns the greeting.

I.2　This Book

For whatever reason, either now or in the future, you are finding yourself in charge of a collection of animals. Perhaps you are a student studying zoo animal technology and hope to one day become a curator or director of a zoological park, aquarium, or sanctuary. Perhaps you are in charge of an animal research laboratory. Maybe you are working at an animal shelter. Maybe you are merely interested in the care of animals and the science of maintaining an animal's welfare. If so, you should find valuable information within this book that should serve you well now and in the future.

This book proceeds from the standpoint that, no matter what the mission of your facility, institution, or industry is, *your* mission as a caretaker of animals is the *welfare* of the individuals you care for and the safety of those working with you. This book should prove relevant whether you are working in a sanctuary, a zoological park, a laboratory, or any other type of facility caring for animals.

For the purposes of this book, all professional animal caretakers, whether Director of a zoological park, entry-level zoo keeper, laboratory technician, or any other job title, are considered 'care technicians'. After all, no matter the facility, the mission of the care technician should be the same. Again, this mission is the welfare of the individuals being cared for.

Supplemental to this overarching mission of animal welfare, are the roles and responsibilities of each care technician. They are as follows:

- *A care technician must offer consistent care.* Implicit in this goal is the fact that care practices must take the form of written protocols that everyone involved in the direct husbandry of an animal, or population of animals, must follow. When different care technicians are following different protocols, the welfare of the animal is compromised.
- *A care technician must make decisions based on objective evidence.* Decisions to alter, add, or subtract anything from a current care protocol must be made based on available data regarding the individual, population, or species. Data regarding an individual or population must be gleaned from assessment, observation, and examination. Data regarding a species must be gleaned from peer-reviewed published sources.
- *A care technician in charge of an animal's care must be educated and properly experienced before undertaking the responsibility.* Experience can be garnered from apprenticeships, internships, or hands-on programmes that provide relevant training. Additionally, a care technician must be educated about the species they are responsible for. Those managing a staff of care technicians must base their hiring practices on this.
- *A care technician not only promotes the welfare of the animal in their care, but also the safety of those working with them.* Providing and maintaining a culture of safety ensures that all staff are following the proper standards and practices; critical to the welfare of the population of animals being cared for.

I.3   Introduction to the Curatio Fundamentorum

Throughout this book, we will be regularly referring to certain fundamentals of care that are central to maintaining the mission of maintaining the welfare of an animal in our care. These stated fundamentals are at the foundation of all operations of animal care and should serve as mandates for all involved in the direct husbandry of an individual or population of animals. These extensive fundamentals are *The Curatio Fundamentorum.*

The Curatio consists of seven major fundamentals. They are as follows:

1. All species are unique, with unique needs, unique challenges, and unique measures of thriving.
2. An individual's welfare is directly correlated with an individual's ability to thrive in a given environment.
3. Captivity, by its definition, removes an individual from a natural ecology, displacing physiological and psychological mechanisms, and thus negatively affects an individual's ability to thrive. As such, artificial environments must compensate for what has been lost and mitigate what has been displaced.
4. An individual's environment is a combination of the following: a habitat; the degree of physical movement possible; the degree of mental stimulation present; the social dynamics with conspecifics; and the degree of intrinsic motivation afforded. The variables of these external factors determine an individual's ability to thrive in an environment.
5. Physiological and psychological mechanisms allow an individual to survive. The health of these internal factors determines an individual's ability to thrive.
6. There is an inherent risk in caring for another species. A culture of safety protects both the caretaker and the cared for.
7. The care of living things is a science and must be approached scientifically.

Complementing the seven major fundamentals are critical sub-fundamentals. All are equally crucial to the care and welfare of animals in managed populations. Each fundamental and sub-fundamental are numbered accordingly, for easy reference.

The Curatio Fundamentorum

1. **All species are unique, with unique needs, unique challenges, and unique measures of thriving.**
2. **Welfare is directly correlated with an individual's ability to thrive in a given environment.**
   2.1. Thriving is defined as a positive physiological, psychological, and ecological state in relation to an individual's physiological, psychological, and ecological needs.
      2.1.1. Measures of thriving are species-specific.

Figure I.4 Elephant at the Kansas City Zoo. (Photo credit: author).

2.1.2. Thriving is equally defined by an individual's internal factors as well as the external factors of a given environment.

    2.1.2.1. Each species have evolved methods of interacting with their external environment. These interactions are a species' ecology.

        2.1.2.1.1. A species' ecology has evolved throughout its natural history and dictates the external, or environmental, needs of the species.

        2.1.2.1.2. The ability to exhibit one's natural ecology is a measure of thriving.

    2.1.2.2. Each species has evolved internal mechanisms in order to survive their given environment. These internal mechanisms are a species physiology and psychology.

        2.1.2.2.1. A species' physiology and psychology have evolved throughout its natural history and dictate the natural behaviour and cognition of the species.

        2.1.2.2.2. The abilities to utilise one's natural physiology and stimulate one's natural psychology are measures of thriving.

    2.1.2.3. In addition to being species-specific, measures of thriving are specific to individuals.

      2.1.2.3.1. Individuals of the same species can have individual challenges and barriers to thriving.

         2.1.2.3.1.1. Individual challenges and barriers to thriving can be physiological and/or psychological.

3. **Captivity, by its definition, removes an individual from a natural ecology, displacing physiological and psychological mechanisms, and thus negatively affects an individual's ability to thrive. As such, artificial environments must compensate for what has been lost and mitigate what has been displaced.**

    3.1. Animal species live in artificial environments for a myriad of reasons by entities with a myriad of missions and objectives. However, the ultimate goal and top priority for those providing husbandry is the welfare of the individuals in their care.

    3.2. Positive animal welfare for animals in human care, on a species level, means that individuals are effectively compensated for the species-specific factors of thriving that have been deprived to the individual by captivity.

    3.3. Positive animal welfare for animals in human care, on an individual level, means that husbandry protocols are tailored to an individual's unique challenges and barriers to thriving; thus, an individual is provided with individualised care.

4. **An individual's environment is a combination of the following: a habitat; the degree of physical movement possible; the degree of mental stimulation present; the social dynamics with conspecifics; and the degree of intrinsic motivation afforded. The variables of these external factors determine an individual's ability to thrive in an environment.**

    4.1. The determining factors of an individual's ability to thrive in an environment are both species-specific and individually specific.

    4.2. An individual living outside of a natural environment is living in an artificial environment.

      4.2.1. All captive environments are artificial environments.

    4.3. The foundation of an environment is a habitat.

      4.3.1. In captivity, an individual lives in an artificial environment.

        4.3.1.1. In order for an individual to thrive in an artificial habitat, the habitat must provide the same ability for physical movement that is provided in a natural habitat.

          4.3.1.1.1. If the same ability for physical movement cannot be achieved in an artificial habitat as in a natural habitat, alternative methods of providing physical movement must be provided to compensate for this loss.

        4.3.1.2. In order for an individual to thrive in an artificial habitat, the opportunity to engage with the features of a habitat must mirror the opportunities for engagement that an individual would find in a natural habitat.

            4.3.1.2.1. If the same opportunity for habitat engagement cannot be provided in an artificial habitat as in a natural habitat, alternative opportunities for habitat engagement must be provided.

   4.4. Enrichment is a process by which artificial stimulation is introduced to an individual or group in captivity in order to compensate for natural stimuli lost to the individual through the course of living in an artificial environment.

      4.4.1. In an artificial environment that promotes thriving, enrichment is regimented as part of a daily care plan.

      4.4.2. In an artificial environment that promotes thriving, enrichment provides psychological stimulation that mirrors the species-specific stimulation that an individual would receive in a natural environment.

      4.4.3. In an artificial environment that promotes thriving, enrichment provides physiological movement that mirrors the species-specific physiological movement that an individual would perform in a natural environment.

      4.4.4. Some individuals may require unique enrichment to mitigate individual challenges to thriving. This is individualised enrichment.

   4.5. The dynamics of the interactions between conspecifics are species-specific and critical to the well-being of an individual. These dynamics are an individual's social environment.

      4.5.1. A rich social environment can be achieved in an artificial environment when the degree of interactions between conspecifics mirrors that of the interactions in a natural environment.

         4.5.1.1. The dynamics of a rich social environment are species-specific. This specificity determines factors such as population size, sex ratio, or any other species-specific variable that defines a natural social environment.

         4.5.1.2. A rich social environment is critical to thriving.

         4.5.1.3. The lack of a rich social environment cannot be effectively compensated.

   4.6. In an environment that promotes thriving, an individual is intrinsically motivated to interact with that environment.

   4.7. Artificial environments must be constantly assessed to ensure that the physiological, psychological, and social needs of all occupying individuals are being met.

5. **Physiological and psychological mechanisms allow an individual to survive. The health of these internal factors determines an individual's ability to thrive.**

   5.1. In a captive environment, an individual's psychological well-being can be monitored.

      5.1.1. Psychological health can be monitored through behavioural observation and assessment of an individual.

         5.1.1.1. Behavioural observation and assessments should be objective.

5.1.1.1.1. Assessments should use predefined indicators of positive and negative psychological well-being.

5.1.1.1.2. Assessments should be designed to produce statistically valid results.

5.1.1.1.3. Assessments should be used to draw a conclusion on the individual's current state of well-being.

5.1.1.1.4. Assessments should be used to craft an individualised care plan for the individual, as well as to make any changes to an artificial environment.

5.2. In a captive environment, an individual's physiological well-being can be monitored.

5.2.1. Physiological health can be monitored through observation, body condition scoring, routine physical examinations, and other methods of assessing an individual's medical health.

5.2.1.1. Medical observation should be routine and performed by a veterinarian with a working knowledge of the species being observed and used to determine an individualised medical care plan.

5.2.1.2. Body condition scoring assesses the physical appearance, weight, or other relevant indicators of medical well-being in an individual. Body condition scoring should be performed by a veterinarian with a working knowledge of the species being scored and used to determine an individualised medical care plan.

5.2.1.3. Physical examination should be done at regular intervals, performed by a veterinarian with a working knowledge of the species being examined, and used to determine an individualised medical care plan.

5.2.1.4. Novel methods of monitoring an individualised physiological health should always be objective, evidence-based, and produce statistically valid results. Results should be used to determine an individualised medical care plan.

5.3. Proper nutrition is achieved through a diet that takes into account species-specific needs as well as individually specific needs.

5.3.1. An individual's diet should provide the same nutritive elements that are found in the species natural diet.

5.3.2. An individual's diet should take into account specific health conditions and dietary needs.

5.3.3. An individual's diet should take into account current body condition.

5.3.4. An individual's diet should take into account an individual's activity level.

5.3.5. An individual's diet should be enriching.

5.3.5.1. A diet should vary.

5.3.5.2. A diet should include noted individual food preferences.

5.3.5.3. A diet can be presented in various and novel manners and forms that support enrichment.

5.4. Medical treatment, including routine treatments, triage, emergency care, and euthanasia, should proceed according to a written plan of veterinary care.

5.4.1. A plan of veterinary care should be authored by a veterinarian with a working knowledge of the captive species.

5.5. Operant conditioning is a process whereby an individual is trained to participate in their own care.

5.5.1. Operant conditioning uses reinforcers to shape and maintain behaviours that can be used for both physiological and psychological care.

5.5.1.1. Positive reinforcers are generally more effective and lasting than negative reinforcers.

5.5.2. Operant conditioning can reduce the stress and anxiety on an individual during care procedures.

5.5.3. Operant conditioning can improve the consistency of care.

5.6. The quality of the environment occupied by an individual is a critical component of psychological and physiological well-being.

5.6.1. Clean water must be provided to all individuals at all times, regardless of species.

5.6.2. Food must be dispersed and contained in a sanitary container.

5.6.3. A clean habitat must be provided to all individuals at all times, regardless of species.

5.6.3.1. Cleaning protocols must be consistent.

5.6.3.2. Cleaning protocols must include a regular disinfection schedule.

5.6.4. Individuals must be provided a habitat with substrate that effectively breaks down waste.

5.6.5. Individuals must be provided with proper bedding, relative to their species.

5.7. When providing care to another species, anthropomorphism is deleterious to maintaining and assessing an individual's physiological and psychological health.

5.7.1. Anthropomorphism seeks to assign human-specific justification to non-human behaviours and other traits.

5.7.1.1. Anthropomorphism must be recognised when it occurs.

5.7.1.2. Anthropomorphism can lead to deep misunderstandings when observing and assessing the condition of a non-human individual.

5.7.1.3. Anthropomorphism must be avoided while assessing the condition of a non-human individual.

6. **There is an inherent risk in caring for another species. A culture of safety protects both the caretaker and the cared for.**

6.1. A culture of safety promotes addressing dangerous situations before they arise.

      6.1.1. A culture of safety provides staff training, emergency drills, and proper attire and equipment to all staff members working in and around an animal care facility.

    6.2. A culture of safety effectively addresses emergency situations as they occur.

      6.2.1. A culture of safety operates according to written emergency response protocols.

        6.2.1.1. Emergency response protocols must be as inclusive as possible for any potential situation that may arise.

        6.2.1.2. Emergency response protocols must address a chain of command and general procedure in the event of an unforeseen emergency situation.

      6.2.2. A culture of safety provides for the evolution of protocols in the aftermath of an emergency situation.

        6.2.2.1. In the aftermath of an emergency situation, an after-action review should be conducted to determine the effectiveness of the emergency response. The results should be used to amend any response that was ineffective.

    6.3. At all times, in captive animal care, individuals must be recognised and respected for the species they are.

      6.3.1. Captive animal care is an interspecies interaction.

        6.3.1.1. Interspecies interactions are unpredictable.

        6.3.1.2. Different species react to different stimuli.

          6.3.1.2.1. That which provokes an individual to react can be species-specific.

          6.3.1.2.2. That which provokes an individual to react can be individually specific.

        6.3.1.3. Different species can have unique reactions to different stimuli.

          6.3.1.3.1. The reaction to any given stimuli can be species-specific.

          6.3.1.3.2. The reaction to any given stimuli can be individually specific.

        6.3.1.4. The reactions of any given species, or any given individual of a species, can present risk and danger of varying degrees to human caretakers.

      6.3.2. In captive animal care, a culture of safety is constructed upon a knowledge of the species, as well as the recognition that all interspecies interactions can be unpredictable.

7. **The care of living things is a science and must be approached scientifically.**

My Own Case Study: Mara

With a rag tied to his head like an old buccaneer, the guard at the defunct Puerto Rican zoo leads me to an opening in a wooden fence. Trying to make the most of my broken

Figure I.5 Mara living alone. (Photo credit: author).

Spanish, he soon realises that I am here to see a singly housed chimpanzee named 'Mara'. He points me to a pathway and explains, as best he can, that I merely need to walk down this path, and it will eventually lead me to the chimpanzee enclosure.

The zoo has been closed for almost six years, when a major hurricane devastated the entire island of Puerto Rico. Since this time, the zoo has never reopened. However, the animals, which include lions, hippos, an African elephant, and this singly housed chimpanzee, have remained. The staff of seven zookeepers make the best of the situation and attempt to care for these animals without the benefit of electricity or running water.

Each step I take through the abandoned zoo, paints a more disturbing reality. The overturned shelters, empty corn crib cages with weeds climbing through the mesh walls, and cracked pavement, all contrast with the sounds one would normally hear in a zoo. Lions roar, exotic birds make their calls, and the melee of the buzz of other undefinable species echoes through the air. The sense of normality that these sounds provide only makes the wounded landscape more disorienting. They signal the unprovided needs that the multitude of animals housed at this zoo are being deprived.

Finally, I reach the chimpanzee enclosure. There, in the middle of the enclosure, seated atop a platform, sits Mara (Figure I.5). She looks at me, only giving the slightest hint that she recognises my presence. In her enclosure there is, what looks to have previously been, a now non-functioning pond and waterfall, It is covered with weeds that have grown up through the cracks in the pavement. Three faux stone walls surround the enclosure; thus, making Mara's only vision of the world outside her enclosure to be that either straight in front of her or directly above her. Even this vision is obstructed by the soft chain-link mesh that blankets the entire space.

Beside Mara's enclosure, there are three capuchin monkeys that Mara can only hear. If she hangs from one part of the soft mesh and peers over the wall, she can catch a glimpse of them if they are also in the right spot. Otherwise, there isn't much else for her to see – just abandoned cages and the occasional passerby staff member.

I am here representing the chimpanzee sanctuary I was working for. I am taking part in the first step of our process to determine if our sanctuary is a good fit for a chimpanzee in need. I am looking for signs that would tell me how likely it is that Mara could thrive in an environment like the sanctuary, which features large island habitats, large social groups (often over 20 individuals), daily cognitive and social enrichment, and the freedom of choice to determine how one spends their day across acres of habitat and with so many social companions. While this may seem good for every chimpanzee, many individuals have challenges to living in this type of environment. Chimpanzees that have been deprived of socialisation with other chimpanzees may not be able to be fully integrated into large social groups. Chimpanzees that are not used to large outdoor habitats may have a hard time adjusting to the environment and adjusting to their new-found relative freedom. Some chimpanzees may have had such severe trauma, that they have such anxiety as to not be able to interact with their environment and need an intensive care programme. None of these issues would preclude a chimpanzee from coming to our sanctuary, but knowledge of this helped us to determine a plan going forward of how we can help this chimpanzee thrive, and if our facility is the best place for her to thrive.

Mara is currently living in a situation that prevents her from living life to its fullest potential. Mara is a highly intelligent, highly inquisitive, and by nature, should be a highly social chimpanzee, who had the unfortunate fate of having the world of captivity thrust upon her at birth – a world where she has been deprived of the ability to realise the hallmark traits of her species and the ability to live life to its fullest as an individual. The circumstances that Mara currently finds herself in are a violation of any definition of animal welfare. Despite all of this, Mara appears resilient. She spends her day exploring the small space she has and observing the small view of the world she is given.

I, however, do not believe that captivity has to mean that individuals are deprived of being able to thrive. I do not believe that Mara's situation needs to be indicative of animals in artificial environments. I believe in the science of animal care and the evidence-based techniques of providing for animal welfare. It is the science that is fostered by a philosophy of care that emboldens this idea. I have always believed that a sanctuary's mission should rest in the notion that, while we may not be able to provide an individual with the world their species occupies in the wild, we can create an environment where, despite a captive situation, we can provide measures that allow them a life where they can exude species-typical traits in a species-typical world while providing a clean, enriching, and interactive environment with access to proper nutrition and medical care. Additionally, we can provide as much freedom of choice and self-determination that the limits of their habitat can allow. In this way, we can utilise science to ensure that every individual in our care can thrive.

I am here, watching Mara, to see if we can provide her with such an environment. I do not believe that Mara needs to be sentenced to such a life. Her life history may have created challenges to being able to currently live life to its fullest potential. However, it is our responsibility to see that she is given the tools, and the environment, to overcome these challenges. Mara deserves a chance to tap into the natural resilience of her species. The science of animal care can provide her that chance.

It is in this spirit that I have set out to write this book. It is written under my firm belief in the foundation of evidence-based, individualised animal care, with individual thriving as its priority. It is written with one prevailing ethic: that no matter the mission of an institution housing captive animals, welfare and the ability for an individual to thrive must always be the prime concern. It is written under the resolution that providing animals with self-determination, a cognitively stimulating environment, a proper social environment, a healthy environment and all in a world where they may thrive, is a science and must be treated as such.

POSTSCRIPT: Mara never came to the sanctuary. Like a lot of things in our field, various complications, politics, and finances came into play. However, I can happily report that a zoo stepped in and was able to bring her in. Mara now lives with a rich social environment, in a highly enriching setting – a place where she can truly thrive.

Part I

Thriving

***Curatio Fundamentorum* 2:** Welfare is directly correlated with an individual's ability to thrive in a given environment.

1   Defining Thriving

Figure 1.1 Penguins at the Saint Louis Zoo. (Photo credit: author).

1.1   Thriving versus Just Surviving

Imagine a newborn human child. As soon as she is born, she is taken from her mother and placed in a white room. There she is confined to a crib. Each day, she is attended by nurses who make sure that she is properly fed, free of disease, and her life is free of incident. She is given no hanging mobiles to watch, no music to hear, no stuffed animals to hold, and a lack of colours to see. Her world ends at the white walls that surround her.

As she proceeds from infancy to toddlerhood, she instinctively begins to sit up. She starts to move around. She shows signs of being driven to explore her surroundings. At this point the nurses take her from her crib and move her to a chair, where she is strapped down. For the greater part of each day, she is left completely alone, confined to this chair. However, each day, the nurses come in, check on her, and move her limbs around so that her muscles don't atrophy. She receives a balanced nutritional meal, three times a day. She is given a daily physical to make sure that she is healthy. She is cleaned up. Her room is sanitised. Each night, she gets moved from the chair, into a bed, where she is also strapped in. No one speaks to her. No one teaches her anything. No one interacts with her except to clinically attend to her basic needs.

As the child grows up, she remains healthy. She is surviving. However, she has been given none of that which humans need to *thrive*. Humans are an incredibly social species. Humans are also a cultural species that rely on teaching and learning to survive the environment around them. As such, humans are born with an inherent drive to interact with other humans. Being a highly cognitive species, humans require the ability to explore their surroundings and the ability to interact with their environment. However, this child has been given none of this. She has been given none of the cognitive stimulation that a human would require. She has been deprived of the rich social environment that is so important for every human to have. She has not been given any sense of self-determination or any freedom of choice that a human would require to realise any sense of their own well-being. It's safe to assume that, even though she doesn't know anything else, she leads a tortuous existence – simply due to the fact that she has been robbed of what she needs to thrive as a human child.

Therein lies the difference between 'surviving' and 'thriving'. This child may be healthy. She might have been given all of her basic requirements for survival. However, what she has been given none of are the opportunities to thrive.

Care technicians must provide opportunities to thrive. If we are to be responsible for the care and well-being of another individual, or group of individuals, it is our responsibility to see to it that those in our care have every opportunity to thrive that can possibly be given to them.

We can define **thriving** by the following definition: It is the constant, spacious pursuit of living life to its fullest potential.[1] In our horrifying scenario, this little girl has been given none of that – even though she's been given all that she needs to just survive.

If we look at animals in captive care, we can measure their well-being as to whether or not they are given the opportunity for that constant conscious pursuit of living life to its fullest potential. Obviously, that pursuit is going to be species-specific. It's going to be specific to what that species has evolved to need and require in order to thrive; in order to live that life to its fullest potential. So what might be true of a human child might not necessarily be true of a lizard, or bird, or even another primate. Therefore, in order for us to truly determine what an animal needs to thrive, a clear and complete understanding of that species is required by everyone in charge of their care and well-being.

---

[1] See Maple & Perdue, 2013 and Maple & Bloomsmith, 2018.

1.2   Aspects of Thriving

Depending on our background, or our individual area of expertise, we may be inclined to focus on just one or two aspects of thriving. Perhaps we are focused on just the physical aspects of thriving – how an individual is acting with regard to their health, whether it appears that they feel okay or appear sick, or what an individual's range of movement is. Perhaps we are focused more on the behavioural aspects of thriving – how social an individual may be, or if an individual is properly engaging with their environment. We may even be mostly focused on the psychological aspects of thriving – looking at signs of psychological distress or psychological well-being. However, focusing on any one aspect of thriving is a mistake. Thriving, by its very nature, is an all-encompassing indicator of general well-being. For this reason, providing opportunities to thrive must be approached *holistically*. This requires that every facet of an animal's direct care team must work together to determine and enable opportunities to thrive; including veterinary staff, husbandry staff, behavioural staff, and facilities staff.

Providing opportunities for animals to thrive is a painstakingly complex operation. Indeed, providing these opportunities is a science. In fact, it is, all at the same time, a physical science, a natural science, and a behavioural science. Therefore, in order for it to be done properly, it must always be viewed as a **holistic** science.

There are two critical methods with regard to allowing animals to thrive in one's care. The first involves *finding opportunities* for animals to thrive. The second involves *mitigating challenges and removing barriers* to thriving for each individual.

The method of finding opportunities for animals to thrive begins with the recognition that a captive environment, no matter how good or well executed, is not a natural environment. As such, a captive environment, unto itself, limits the opportunities for individuals to thrive. Therefore it is incumbent on each care technician, whether they are involved in husbandry, veterinary care, behavioural analysis, or facility design and upkeep, to find ways to provide opportunities, even in captivity, for individuals to thrive. These general opportunities may include providing individuals with the tools they need (the tools they would have in nature) to thrive. This might include adequate space, appropriate features within a habitat, a care plan that features a natural daily activity budget, a proper diet, or any other aspect of care. All in all, providing opportunities for animals to thrive is something that must be carefully planned and in accordance with which species one is taking care of.

The second method, mitigating challenges and removing barriers to thriving, operates under the recognition that every individual animal has their own unique sets of challenges to being able to fully thrive. Even under the best care plans, even with the best facility design, even with the best veterinary care, individuals may still struggle to thrive due to their own unique circumstances. These circumstances might be medical. They may be psychological. They may be behavioural. They may even be a mix. It is incumbent on each care technician to recognise what these challenges are for each individual in their care and attempt to mitigate these barriers to thriving. Once again, it takes a holistic approach where every care technician involved in any aspect of direct care, must be involved in both recognising these challenges and finding ways of mitigating these challenges.

Physical Thriving

An individual's physical state is foundational to one's ability to thrive. Illness, injury, or physical disability can greatly impair an individual's ability to live life to the fullest. All individuals have challenges, varying degrees and types, to physical thriving. These challenges can be mitigated, or even prevented, through proper care and management.

The ability for physical thriving begins with the direct husbandry of an animal. Cleaning protocols, feeding protocols, enrichment, operant conditioning, and general care plans must be designed to keep an individual healthy. Proper movement must be encouraged in care plans. The way animals are fed must be species-specific and allow individuals to eat as they have evolved to do. The planning of enrichment must be tailored to the species-specific needs of that individual to cognitively stimulate them properly, promote proper movement, promote necessary socialisation, and to allow the proper engagement with their environment (Figure 1.2). A clean environment where animals live and eat must be maintained at all times. Husbandry staff can help keep animals healthy by storing food items, enrichment items, and anything else that might come in contact with an animal, properly and sanitarily.

Obviously, good veterinary care is essential to physical thriving. In addition to the proper treatment of ailments and injuries, a good veterinary programme practises preventative care. A good veterinary programme understands each individual's baselines through routine physicals, body condition scores, quality of life assessments, and detailed record-keeping. Such a programme makes objective and evidence-based conclusions of each individual's health and physical well-being and determines plans on how to improve that physical well-being.

A robust animal behaviour programme is also key to this goal of providing opportunities for individuals to physically thrive. Routine behavioural welfare analyses should be performed on every animal to determine where an individual may be facing physical issues. These assessments may score basic indicators of physical welfare and

Figure 1.2 Griffon vulture at the Suffolk Owl Sanctuary. (Photo credit: author).

can be used in collaboration with a veterinarian and used to determine if these individuals are behaving in such a way as to create concerns that they may not be physically thriving. Are they acting in such a way that indicates a malady?

Finally, the facility's design and upkeep are immensely important in an animal's ability to physically thrive. Enclosures must be designed to allow for the species-specific physical mobility for each animal. When populations age, the necessary adjustments must be designed and executed according to what that particular species physically requires during geriatric years. Physical disabilities must be addressed at the facility level by designing structures within an enclosure that can help mitigate these challenges so that every individual has the ability to live and engage in normal activities for that species as much as possible.

Dimensions of Physical Thriving

* *The ability for a natural range of movement*

    Each species has evolved a way to move throughout their natural habitat that allows for them to survive. In turn, each species has evolved physical characteristics that allow for their specific **range of movement**. Each species is naturally inclined to utilise these physical characteristics in the manner in which they have evolved to use them. In captive situations every effort must be made to allow for this. Artificial habitats must be designed with this in mind. The ability to exercise movement in the way one is naturally inclined to do is key to physical thriving and foundational to welfare. Understanding an animal's natural ranging patterns, natural physical movements, and the ways an animal uses each extremity, is key to providing a proper range of movement that allows each individual to thrive. For example, a bird of prey needs the ability to fly. A sea lion requires the ability to both swim and climb out of the water.

    Let's take the case of gibbons in an artificial environment. A gibbon is a small ape that is often seen housed at zoos. Gibbons have an extremely unique morphology that is based on a unique range of motion. In the wild, gibbons live in dense forests. They are rarely, if ever, observed on the ground. Instead, they have evolved a method of locomotion called brachiating whereby they move from tree limb to tree limb very rapidly. In order to do this, they have evolved very long fingers, long arms, and short legs. In order to properly house gibbons in an artificial environment, they must be given structures that allow for an arboreal lifestyle. When they are not provided such structures, gibbons have a very difficult time moving around their habitat – having to walk bipedally while holding their long arms in the air.

    The consequences of not allowing a natural range of movement are profound. Animals without this ability can suffer from such maladies as muscle atrophy, circulatory problems, psychological issues, and more.[2] However, with a proper facility design, care plans, and management protocols, many of the challenges to a proper range of movement can be mitigated.

---

[2] See Hediger, 1950.

- *The ability to live free of pain and physical distress*

    Foundational to good general well-being is just feeling good. All aspects of thriving proceed from this. If an individual is in pain, if an individual is feeling sick, or if an individual is, in some way, in physical distress, all other aspects of thriving are compromised. It is impossible to engage, in any normal sense, with the world around you if you are physically compromised. As such, ensuring that each animal in one's care has the ability to live free of physical distress is analogous with good care. When such distress cannot be avoided, such as with chronic conditions, illness, or injuries, all efforts must be made to mitigate the distress and allow the animal to be as comfortable as possible. It is also incumbent on every care technician to practise protocols that would *prevent* physical distress by providing the essentials to good health: good nutrition, a clean and sanitary environment, and a safe environment.

    Providing each animal the ability to live free of pain and physical distress has two components:

    - Maintaining good health while preventing physical distress
    - Mitigating existing conditions.

- *The ability to live according to a species-specific circadian rhythm*

    An **ecological time niche** refers to the manner in which a species has evolved to behave at any given moment during a 24-hour cycle. Or, in other words, a species' natural **circadian rhythms**. Circadian rhythms relate to various physical, behavioural, and mental processes that a species undergoes during this 24-hour cycle. This governs sleeping patterns, feeding patterns, social patterns, ranging patterns, and anything else that an animal needs to accomplish during each day. The ability for a species to practise their natural circadian rhythms is extremely important for good welfare.[3]

    For example, nocturnal species need to be allowed to sleep during the day and be awake during the night, while practising all of their waking behaviours during this time. A spotted owl has evolved all the behaviours that go along with nocturnal living. They need to be able to feed at night. They need to be able to move about at night. They need to be able to have a place that is conducive to sleeping during the day. On the other hand, diurnal species need the opposite. A giraffe, for example, needs to be able to eat, socialise, and roam around during the day, while having a place that is conducive for sleeping during the night. Some species may not fit into either category. For example, lemurs are crepuscular animals, meaning they are most active at dusk and dawn. Therefore, they must be able to eat at dusk and dawn and be able to rest during other times of the day.

    The consequences of not allowing a species to exhibit their natural circadian rhythms can be both physical and psychological. Forcing a schedule on an animal whereby they are fed when they are not naturally inclined to do so can lead to malnourishment or improper nutrition. Not allowing them to sleep when they are naturally inclined to do so can lead to massive sleep deprivation and all of the physical and psychological effects that go along with it.

---

[3] See Berger, 2011.

It is therefore incumbent on every care technician to research the circadian rhythms of the species they take care of, and manage that species in a manner that is conducive to accommodating these cycles. It is also critical to determine if these circadian rhythms can be provided and, if not, make a decision as to whether or not it is proper to keep that species in such an environment.

- *The ability to live free of contaminants and waste*

  There are three key factors in providing a clean and sanitary environment for captive animals:
  - Maintaining a proper cleaning schedule.
  - Maintaining a proper disinfection schedule.
  - Providing a proper substrate that absorbs, breaks down, and controls the spread of wastes.

A sound cleaning schedule is a strategy that involves determining how often an enclosure is cleaned and what that cleaning entails. This is going to be determined by both which type of species is being housed and by how many individuals are living in an enclosure. It also requires determining which tools are necessary for cleaning, and what cleaning agents are required. Once a cleaning schedule is arrived at, it must be maintained consistently.

A disinfection schedule is also critical. It may be necessary to disinfect the enclosure every time it is cleaned. Or it may only be necessary to disinfect periodically. Disinfection typically involves using chemicals that kill bacteria. Depending on the caustic nature of these chemicals, as well as the environmental impact of these chemicals, disposal, and safe handling must be taken into consideration.

Finally, a proper **substrate** is a critical piece in keeping an enclosure sanitary. Substrate absorbs, breaks down, and controls the spread of waste, until such time as an enclosure can be cleaned. In some cases, substrate may be so effective as to only require minimal cleaning every day; such as in the case with **deep litter substrate**, whereby an enclosure is designed to hold several thick layers of substrate that break down wastes so effectively as to only require spot cleaning alongside a periodic complete replacement schedule. In other cases, substrate must be partially or completely removed each time an enclosure is cleaned. Though substrate may require a significant expense, it is no less critical than regular cleaning.

Psychological Thriving

Meaningful engagement with the world around us requires psychological well-being. In turn, psychological well-being requires that individuals be able to fully realise their **range of mental processes**, or the array of mechanisms by which a species has evolved to cognitively process the environment around them. Examples of such mechanisms include: choice, control, and decision-making; exploration; utilising all senses in a meaningful way; solving species-typical challenges; or any other way that an individual may psychologically engage with their environment.[4]

---

[4] See Washburn, 2015.

Figure 1.3 Chimpanzees at Lion Country Safari. (Photo credit: author).

Providing a place where animals can psychologically thrive, requires as much of a holistic approach as anything else (Figure 1.3). Facilities must be designed to allow for proper engagement with the environment that is species-specific. A robust enrichment programme must be planned and executed that provides cognitive stimulation, the ability to solve challenges, and the ability to perceive all manner of sensory data. Animal management protocols should allow as much choice and self-determination as possible given the confines of captivity. Animals must be fed in a species-specific fashion. Animal assessments must look at psychological well-being as an overall factor of animal welfare.

One of the most powerful tools at a care technician's disposal is an operant conditioning programme. Operant conditioning, when it takes the form of **positive reinforcement training** (often abbreviated as PRT), allows an animal to voluntarily participate in their own care. PRT involves shaping and maintaining behaviours by offering an immediate reward for desired behaviours. This can be used for veterinary procedures. This can also be used for daily husbandry protocols. This can even be used as an enrichment. For the animals, the process is completely voluntary and is highly stimulating. It can go a long way in eliminating some of the more stressful tactics in animal husbandry, such as the use of restraints, using dart guns, or anything else that might add undue stress or anxiety to an individual. Therefore, a robust operant conditioning programme is critical for any good captive animal management plan.

Dimensions of Psychological Thriving

- *The ability for self-determination*

    Whether by instinct or learnt behaviour, all animals make choices throughout their lives, and at any given time of their day. The ability to make choices are how animals engage with the world around them, and survive the world around them. Depriving an animal the ability to make choices is one of the ways captivity can be such a negative experience. In fact, the very definition of captivity means to deprive self-determination and is based on the lack of ability to make one's own choices. Movements are managed. Behaviours are managed. The ability to make choices is stymied.

    However, as care technicians, we often find that there are areas of captivity where we can offer the ability for choice and self-determination to improve the general welfare of the animals in our care. This may involve providing enough space for animals to move around more freely. This may involve minimising the amount of time that animal movements are managed; operating husbandry protocols based on where animals choose to be rather than forcing them to shift into another area. This may involve designing enrichments where making choices is a key feature. All in all, providing the ability for as much choice and self-determination as possible should be a mantra for all care technicians and all animal husbandry programmes.

- *The ability for psychological stimulation, dictated by species-typical actions and behaviours*

    An animal management programme must provide regular psychological stimulation throughout all aspects of an animal's environment and daily activities. Key to this is for this programme to elicit species-typical actions and behaviours (e.g., an enrichment programme should encourage the psychological and physical processes that an individual would typically encounter in the wild). Accomplishing this requires a psychologically stimulating habitat, feeding protocols that mirror foraging in the wild, providing the ability for an animal to practise natural nesting/bedding/burrowing behaviours (when applicable), allowing for a species-typical social environment, and a robust enrichment programme.

    An enrichment programme is necessary for providing the ability for psychological stimulation. Enrichment should be designed to allow for animals to solve species-typical challenges, while receiving species-typical rewards. This is not to say that enrichments have to mirror what the animal encounters in the wild. However, if it is to be effective, it should encourage similar processes that an animal would encounter in the wild. For example, an African elephant may be given an enrichment where a preferred food source is encased in a block of ice. The elephant must work at getting through the ice in order to reach the reward. An African elephant is never going to encounter ice in their natural habitat. However, they are tasked in the wild with having to figure out ways to reach a food source. The enrichment is effective because it offers psychological stimulation in the form of a challenge that allows the elephant to solve a task through a similar mental process that would be required in the wild.

Similarly, facilities can be designed that offer animals species-typical psychological stimulation. For example, giraffe feeders can be designed to allow giraffes to browse as they would in the wild. Flamingo feeders can be designed to allow flamingos to find food in water, and strain the food through their bills as they would in the wild. Orangutan habitats can be created to provide ever changing arboreal pathways for orangutans to travel around.

- *The ability to engage with one's environment in a species-typical manner*

An animal may be provided with a very large, lush, and beautiful habitat. However, if there is no ability for the animal to engage with that environment, it is meaningless. For example, if arboreal primates live in a large enclosure, but do not have adequate climbing structures, the ability for them to engage with their environment is extremely limited. If a group of meerkats are given a large valley to live in, but there is no ground suitable for burrowing, there are significant challenges to their potential to engage with their environment.

Engagement, like most other aspects of psychological thriving, can be promoted by enrichment and facility design. However, individuals may have unique challenges to being able to engage with their environment. Likewise, individuals must be regularly assessed to determine their level of engagement. When it is concluded that an individual is not regularly engaging with their environment, efforts must be made to determine the cause of the lack of engagement (is it due to a deficiency in facility design? The enrichment programme? Does this individual have a unique challenge or barrier to being able to properly engage with their environment?). If it is an individual challenge, a care plan must be created in order to mitigate the issue.

- *The ability for sensory stimulation*

Everything we know about the world is what we have perceived through our senses. Our memories of the past, our perception of the present, and our hopes and fears for the future are based on what we have sensed in our lives. Each time we perceive something, our senses are stimulated. This is the case with other animals as well. However, different species have unique reliance on different senses. Take, for example, wet-nosed mammals versus dry-nosed mammals. Mammals with wet noses have nasal receptors on the outside of their nose. As soon as they enter an area, they can perceive an array of information about the area they are in based on their sense of smell – who else might be in the vicinity, where objects are located, any potential food sources, potential dangers, and a host of other bits of information. They can navigate their environment extremely deftly based on this sense. Contrast this with a dry-nosed mammal. In order for them to perceive information in their nasal receptors, they must inhale through their nose. Once they do this, they can perceive smells and draw conclusions from this perception. However, they are not nearly as attuned to this information as wet-nosed mammals. Instead, most dry-nosed mammals have compensated for this by their sense of vision. They have evolved complex visual perception – depth perception, colour vision, etc. While both examples have both a visual sense and an olfactory sense, the level of perception is different. Therefore, both have unique sensory needs within their environment.

Ensuring that animals have sensory stimulation is critical for psychological thriving. In order to achieve proper sensory stimulation, it is key to know the level of how that species relies on that sense. For example, a very visual species in a dull, visually lacklustre enclosure is not receiving proper sensory stimulation. Similarly, a very olfactory-centred species in an environment where there is no olfactory stimulation is not receiving the kind of sensory perception they rely on.

Sensory stimulation can be achieved through enrichment that features stimulating tastes, sights, smells, sounds, and things to touch. It can also be achieved through an enclosure that frequently changes. Structures can be changed, new items can be added, new sounds can be produced, etc.

Emotional Thriving

While psychological thriving involves mental well-being and the ability of an individual to interact with the world in a fulfilling and beneficial manner, emotional thriving involves managing emotions, such as excitement, frustration, and fear in a productive manner in line with species-typical behaviour.

Let's imagine two different chimpanzees. The first chimpanzee is thriving emotionally. The second displays noticeable challenges. For example, the first may react to a stressful or frustrating situation with a display that brings forth social reassurances from other chimpanzees. The second may react with stereotypic behaviour, such as rocking back and forth; or even self-injurious behaviour such as hair plucking or picking at their skin. For the second chimpanzee, the inability to manage emotional stress would greatly hamper their general well-being.

Through individual care plans, care technicians can promote emotional thriving. Once again, these plans encompass husbandry techniques, veterinary techniques, behavioural analyses, and facility planning and upkeep. Through careful collaboration, care technicians can provide an environment that is both conducive to emotional thriving and identifies and mitigates barriers to emotional thriving.

Of all the aspects of thriving, perhaps the most individualised is emotional thriving. Much of the degree that an individual can thrive emotionally is based on their own unique history. Periods of trauma, peace, moments of good welfare, and moments of poor welfare, can all dictate how equipped an individual is to emotionally thrive. This means that a large portion of a care technician's responsibility in allowing for an individual to thrive emotionally is steeped in mitigating the unique challenges, unique barriers, and unique circumstances that each individual has. This is not to say, however, that mitigation is the only way that a care technician can promote emotional thriving. In fact, at the group and species level, a care technician is responsible for ensuring that animals do not go through moments of undue stress and anxiety; and that every effort is made to eliminate times of emotional distress. A care technician is also responsible for ensuring that the variables that allow for consolation and emotional relief are in place. Once again, this requires a knowledge of species-typical behaviour, as well as a knowledge of an individual's baseline behaviour.

Dimensions of Emotional Thriving

- *The ability to live free of undue stress or anxiety*

    Captivity, by its very nature, can be extremely emotionally stressful for each individual. An animal in an artificial environment is reliant on human caretakers to provide their life functions. Human caretakers provide food, medical care, and oftentimes dictate where an animal is and what an animal is doing. Such lack of control can be extraordinarily stressful.

    Once again, and as with dimensions of psychological thriving, giving individuals as much choice, self-determination, and control as possible, can greatly reduce undue stress and anxiety. When choice and control are not options, these times should be routine and done so in a manner that is as least disruptive as possible. Operant conditioning, once again, can be an extremely useful tool in combating undue anxiety and stress. When animals can voluntarily participate in their own care, the stress and anxiety that normally goes along with living in an artificial environment, can be greatly reduced.

    Regular care and welfare assessments are a critical factor in finding methods of reducing stress and anxiety. Only through these assessments can a care technician make the determination as to what is causing the anxiety and stress with an individual, or group of individuals, and make the necessary changes to their management programme. For example, this is keenly important for animals that are on display that must deal with the stress and anxiety of being in front of unfamiliar human onlookers. This is also extremely important when animals may be housed next to other groups of animals (either of the same species or of different species) that may be causing stress and anxiety. Whatever the case, it is the responsibility of the care technician to remove the animal from the situation that is causing such distress and find a solution where emotional thriving can be promoted.

- *The ability to experience and benefit from anticipatory behaviour*

    Anticipatory behaviour is the act of perceiving something within, or just outside, one's environment and anticipating what comes next. For example, animals may see care technicians arrive in the area with food and show signs of anticipating eating. Animals may also see something that frightens them and anticipate the situation by getting themselves to a perceived level of safety. In both cases, anticipatory behaviour is a positive emotional response. Anticipatory behaviour shows that an animal is interacting and engaging with their environment.

    In order for healthier anticipatory behaviour to occur, animals need to be allowed to sense what is going on around them. It is critical that animals not be surprised by what a care technician is doing. Management plans should avoid any element of 'sneaking up' on an animal in order to administer care. Avoiding surprises can go a long way in diminishing undue stress and anxiety brought upon by husbandry protocols.

- *The ability to be treated for emotional distress*

    Treatments for emotional distress can include: individualised care plans, new enrichment protocols, new operant conditioning plans, and even include behavioural medication. However, before embarking on a treatment for emotional distress, it is imperative to assess an individual's emotional well-being and make a diagnosis of the condition. Once this occurs, it becomes the holistic domain of the veterinary staff, behavioural staff, and husbandry staff, all working together to come up with a solution to treat the emotional distress.

In order to properly assess an emotional issue, a scientifically sound assessment period must occur that gathers enough data and includes a proper analysis of that data to draw a sound evidence-based conclusion. Emotional distress should never be diagnosed or treated with hunches or anecdotal information. In extreme instances, a temporary method can be utilised to deal with the emotional distress. However, this should be used as an interim method only until enough data and a proper assessment can be carried out.

Social Thriving

Key to an animal's ecology is how they have evolved to interact with conspecifics. Some species live in large groups with multiple males and multiple females. Some may live with one female and multiple males. Some may live with one male and multiple females. Some may live in pairs. Some may be entirely solitary. The ability to live in the social environment that one is naturally inclined to live in is the degree to which that individual is socially thriving.

Social thriving doesn't just refer to the population of a group of conspecifics; or, the sex ratio of that group of conspecifics. It also refers to the ability to socialise in a manner consistent with how that species has evolved to interact. For example, a chimpanzee lives in what is called a fission-fusion social environment, whereby a large population of multiple males and multiple females is governed by one alpha chimpanzee. However, this group fissions off into smaller subgroups, with its own dominance hierarchy, but then comes together under the leadership of the alpha chimpanzee. Added to the fact that chimpanzees practise fission-fusion socialisation, they also have a complex cycle of conflict leading to reconciliation and grooming that further coalesces the group. For a group of chimpanzees to be socially thriving, they need to be given such a rich social environment as to be able to live in a fission-fusion social grouping with the ability for conflict leading to reconciliation and grooming. Conversely, there are species of reptiles that are completely solitary. Allowing any more than minimal interactions with conspecifics can lead to undue aggression and undue stress and anxiety. Thus, in order to promote social thriving in animals, care technicians must completely understand the social proclivities of the animals they are caring for.[5]

In the wild, there are an array of purposes for social interactions between conspecifics. For some animals, these purposes can be vastly complex. For example, there may be social transmission of learnt behaviours that enable an individual to survive

---

[5] See Powell, 2010.

their environment.[6] There may be complex communication systems; such as alarm calls that specify which predator is in the area and instinctually warn conspecifics of where to flee from the danger. Conspecifics may interact in an effort to find and obtain food sources. However, at its most basic, social interactions are used for the following: procreation, parenting, and supporting others in times of danger.

Dimensions of Social Thriving

- *The ability to live in a social environment similar to species-typical socialisation in the wild*

  Because animal species have evolved to survive by the manner and degree of their social interactions with conspecifics, it is critical to an individual's well-being that they be given the same types of opportunities to socialise in an artificial environment that they would in a natural environment. Depriving animals of this ability can lead to emotional, psychological, and physical issues. If species that normally live in robust environments in the wild, are forced into a very limited social environment in artificial environments, they can become bored, anxious, or, in severe cases, develop psychosis.[7] For species that normally live in a solitary environment in the wild that are paired with others in artificial environments, they can face significant, and sometimes lethal, aggression. It is therefore necessary for the artificial social environment to mirror the natural social environment whenever possible.

  This is not to say that there are no major challenges that a care technician can face when providing such a social environment to the animals in their care. For example, for species that are both highly social and **transient** (individuals frequently move between social groups), or highly social and **sexually philopatric** (one sex disperses from a social group while the other sex remains within the group), it is going to be very difficult to mirror what occurs in the wild. For situations such as these, care technicians must develop novel methods and care plans to address the shortcomings brought upon by captivity.

- *The ability to perform procreative actions*

  Reproductive and courtship behaviour is an evolutionary-driven need for all sexually reproductive animals. Whether or not a programme or facility allows for captive breeding, animals should be given the ability for procreative actions. However, by providing this ability, care technicians must carefully manage the population to control reproduction (by preventing it in facilities where breeding is not desired, or managing it in facilities where breeding is allowed). Controlling reproduction requires a birth control method, either by temporary measures (daily medications, implants) or permanent methods (surgical). Depending on the mission of the animal care programme, this will obviously vary. However, if animals are able to reproduce, careful measures need to be enacted to both prevent inbreeding and to ensure the safety and health of future offspring.

---

[6] See Hopper, 2021.
[7] See, for example, Ridley & Baker, 1982.

- *The ability to be parented in a species-typical fashion*

When animals are born in artificial environments, every effort should be made, whenever possible, to keep the offspring with the parent. Unless there are behavioural or physical conditions that prevent natural parenting, biological parents in artificial environments (or alloparents of the same species) will almost always provide more of what an offspring needs than a human caretaker can. Depriving an offspring of being able to be parented by their own species can lead to either immediate or future psychological, emotional, or physical issues. Removing an offspring from a parent can be extremely stressful and can similarly cause psychological, emotional, or physical issues.

However, a captive situation, no matter how well executed, is different from a natural situation; and, despite providing the ability for natural parenting, care technicians must be involved. Parents and newborn offspring must be carefully monitored to ensure the health of the offspring and the safety of the offspring. Offspring born into large social groups may face aggression from conspecifics (even their own parents). Offspring born into solitary social environments can be born to unfit parents. Additionally, the health of the newborn must be consistently monitored to determine if the newborn is receiving everything they need from their biological parents and the group.

1.3 Vicino and Miller's Five Opportunities to Thrive

The International Council of Ethologists hosts periodic, and informal, conferences on animal behaviour where experts in the field can present novel ideas on a diversity of topics related to the field. The conference highlights integrative approaches and, because of this, has been host to some groundbreaking ideas.

In 2015, in Cairns, Australia, one such groundbreaking presentation was given by Greg Vicino, of the San Diego Zoo, and Lance Miller, of the Brookfield Zoo in Chicago. The presentation was titled, 'From Prevention of Cruelty to Optimising Welfare: Opportunities to Thrive' and focused on expanding how animal care facilities assess well-being by determining an animal's opportunities to live a life of thriving.[8]

Vicino and Miller centred their presentation around five opportunities to thrive. These opportunities were an expansion on the idea of '**The Five Freedoms**', which is an internationally recognised list of five aspects of animal welfare for animals in captivity. The five freedoms are:

1. Freedom from hunger or thirst.
2. Freedom from discomfort.
3. Freedom from pain, injury or disease.
4. Freedom to express (most) normal behaviours.
5. Freedom from fear and distress.

---

[8] See Vicino & Miller, 2015.

While the Five Freedoms are essential critical foundations of good welfare, they focus on negative states of welfare and do not adequately measure *positive* states. If a care technician is only measuring negative indicators of welfare (for example, self-injurious behaviour as an indicator of fear and distress), then they may be given a false sense that the absence of these indicators signifies good welfare, when, in fact, the absence of a *positive* indicator may be more revealing. Instead, in order to properly assess well-being, the interplay between the positive states and negative states must be measured. In order to achieve this, Vicino and Miller presented their five opportunities to thrive as a complement to the Five Freedoms, focusing on thriving as a measure of good well-being.[9]

The five opportunities to thrive are:

1. *The opportunity for a thoughtfully presented, well-balanced diet*

   Animals require a well-balanced and nutritious diet. This is key to physical thriving. However, in addition to a nutritious diet, animals need food to be presented in a manner that gives them an opportunity to exhibit their natural feeding behaviour and is enriching to eat.

2. *The opportunity to self-maintain*

   Self-maintenance means that an animal has the opportunity to house themselves in appropriate shelters, has the ability to make their own bedding, has the ability to find their own food, and has the ability to enrich themselves. The opportunity to self-maintain is cognitively and behaviourally stimulating.

3. *The opportunity for optimal health*

   Optimal health needs to be assessed alongside poor health. While an animal may not be showing clear signs of illness, they still may not be in optimal health.

4. *The opportunity to express species-typical behaviour*

   Allowing an animal to express species-typical behaviour is accomplished through proper facility design, a sound management programme, behavioural assessments and care planning, and, most importantly, a care technician's knowledge about the species in their care.

5. *The opportunity for choice and control*

   All animals are endowed with the ability to make choices and have control of their behaviours. Choice and control are primary cornerstones of good well-being.

Implications

In our exploration of the definition of thriving, we come to the conclusion that much of what amounts to thriving is species-specific, demands a wealth of knowledge of the species, and must be endeavoured at all costs. Adding to this is the fact that the very definition of thriving may be specific to an individual, based on one's unique circumstances and histories that present specific challenges to thriving.

---

[9] See Miller et al., 2020.

Care technicians are responsible for more than just the life functions of the animals in their care. They are responsible for all avenues of thriving that are possible in a given situation. Giving animals the opportunity to thrive requires comprehensive knowledge of a given species, individualised care, and evidence-based care.

References

Berger, A., 2011. Activity patterns, chronobiology and the assessment of stress and welfare in zoo and wild animals. *International Zoo Yearbook, 45*(1), pp. 80–90.

Hediger, H., 1950. *Wild animals in captivity.* London: Butterworth-Heinemann.

Hopper, L. M., 2021. Leveraging social learning to enhance captive animal care and welfare. *Journal of Zoological and Botanical Gardens, 2*(1), pp. 21–40.

Maple, T. L. and Perdue, B. M., 2013. *Zoo animal welfare* (Vol. 14). Berlin: Springer.

Maple, T. L. and Bloomsmith, M. A., 2018. Introduction: the science and practice of optimal animal welfare. *Behavioural Processes, 156*, p. 1.

Miller, L. J., Vicino, G. A., Sheftel, J. and Lauderdale, L. K., 2020. Behavioral diversity as a potential indicator of positive animal welfare. *Animals, 10*(7), p. 1211.

Powell, D. M., 2010. A framework for introduction and socialization processes for mammals. In D. G. Kleiman, K. V. Thompson and C. K. Baer (eds.), *Wild mammals in captivity: principles and techniques for zoo management* (Vol. 2) 2nd ed. Chicago: University of Chicago Press, pp. 49–51.

Ridley, R. M. and Baker, H. F., 1982. Stereotypy in monkeys and humans. *Psychological Medicine, 12*(1), pp. 61–72.

Vicino, G. and Miller, L. J., 2015, September. From prevention of cruelty to optimising welfare: Opportunities to thrive. In *Proceedings of the International Ethological Conference, Cairns, Australia* (Vol. 19).

Washburn, D. A., 2015. The four Cs of psychological wellbeing: Lessons from three decades of computer-based environmental enrichment. *Animal Behavior and Cognition, 2*(3), pp. 218–232.

2 The Species-Specifics of Thriving

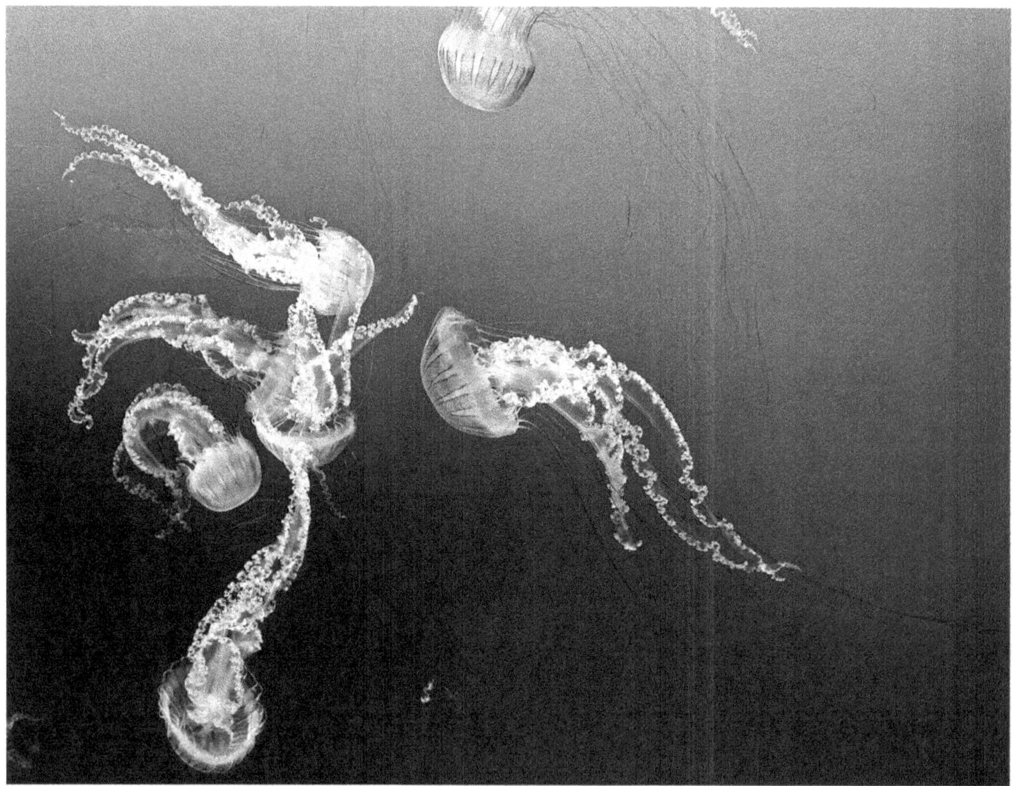

Figure 2.1 Jellyfish at the Kansas City Zoo Aquarium. (Photo credit: author).

2.1   The Ancient Seeds of Thriving

Imagine travelling back in time to about 65 million years ago. Planet Earth is a very different place than it is today. A cataclysmic event has just occurred. The entire planet has become somewhat of a wasteland. The animals that have been dominating the planet – very large cold-blooded animals, are dying out. What is replacing them are animals that have a certain amount of resilience to this new phase of the world.

Some of these animals can regulate their own body temperature, allowing them to survive in a variety of climates. Some of these animals have replaced their scales with feathers to keep them warm in colder climates. Some of these animals have developed mammary glands that allow them to feed their young. These critical moments in the natural history of the Earth have dictated which species have been able to thrive.

Beyond these very major things, there are some more specific things. For example, some of those mammals began to be structured to live a life in the trees. They develop grasping hands and feet that can cling to tree branches. They develop a different limb structure that can manoeuvre them in the trees. They develop a keen sense of vision that allows them to hunt the very small prey that live in the trees. They are gifted with extreme depth perception, stereoscopic vision – where their eyes are forward facing and focused on the same area. These, of course, are the primates. These features of the primate order allow them to thrive in the treetops in the Earth, as it was, tens of millions of years ago.

However, some of these features carry with them a cost. For example, with forward facing eyes and stereoscopic vision, primates lose the peripherals enjoyed by most of the other mammals. This makes them much more vulnerable to predators silently sneaking up on them from behind. In turn, this necessitates another development. Primates have large and complex social groups that can warn each other of danger. This means that primates have had to develop communication systems where conspecifics can warn each other of the presence of danger. Thus, primates have developed genetic alarm calls, social hierarchies, and group coalitions. All of this initially allows primates to thrive in the trees, with extreme depth perception, and stay safe from predators.

Flash forward to the present day. As humans, we are the product of these moments in natural history. Our body type bears the imprint of when our order evolved to thrive and be fit for an arboreal lifestyle. Our ability to navigate our world through our depth perception; such as our ability to drive a car, understand photographs, create and operate machines, create works of art, utilise maps, and every other function based on our ability for stereoscopic vision; has all sprung from the variables of our natural history as part of the primate order, and developed even more specifically in our natural history as the human species. Additionally, our social organisation as a species, our societies, our ability to interact, our ability to teach, our ability to learn, our cultures all stem from the necessary population dynamics that arose out of a compensation, in part, from a lack of peripherals lost to us as a cost of stereoscopic vision. These are just some of the millions of aspects of being human based on our unique natural history. Each one of these aspects is absolutely critical for us to thrive.

Every species on earth, every taxonomic group, has specifics based on unique natural histories that have allowed them to thrive in the natural world. This must be reproduced in artificial environments for individuals of that species to thrive in any capacity.

2.2 Selection

Each species on the planet exists as a collection of physical and behavioural traits. This unique assembly defines each individual species. It is a reflection of its natural history and a reflection of each ecological pressure that the species has undergone over the course of its existence. In fact, each species can be viewed as individual stories of natural history. Each ecological pressure that a species has encountered has dictated the selection of a physical or behavioural trait that has allowed that species to compete and thrive in the given environment. Each successive ecological pressure has determined the selection of new traits; each new trait having to build on previous traits; in turn, shaping how the new trait mitigates the new ecological pressure.

For example, a terrestrial primate became ground dwelling, rather than tree dwelling, based on some ecological pressure encountered by the species at some point in its natural history. However, though they selected terrestrial traits, their physicality bears the shadows of a previously arboreal time (limb structure, visual apparatus, balance, etc.). The mixture of these shadows of the past along with presently selected traits is what creates the unique species as it is. It is in this way that species have become specialised to their own unique environment. This is how species realise their niche. Outside of that environment, they cannot realise their niche, and, as such, cannot thrive with their existing traits. Thus, their general well-being is compromised.

Darwin's famous finches are examples of this. A finch with a long thin beak only selects that trait if they live in an area where they can outcompete others if they live in an area where there is a food source that can be most easily obtained with a long thin beak, such as food that lies in tiny crevices. Conversely, finches with short, fat beaks can only thrive in areas where food sources are best obtained with their unique beaks, such as food with hard shells. If one were to force the two finches to swap ecosystems, they would not be able to realise their niche, outcompete other species for food, and thrive.

It is for this precise reason that artificial captive environments must do their best to mirror the natural ecosystems that the species would encounter in the wild. These captive environments must allow each species to utilise both their physical and behavioural traits in a way that allows them to mirror their niche they would encounter in a natural ecosystem; the niche that both determines and utilises their traits as a species. Anything short of this, can cause a severe compromise of animal welfare.

Speciation-Level Pressures

Ecological pressures lead to evolutionary adaptations that, over time, can create major changes in a population. When changes reach a certain level of significance, speciation can occur. **Speciation** is when a new species is defined by its unique evolutionary changes from an original parent population. These traits that have signified speciation are the **hallmark traits** of a species. These hallmark traits, along with the traits that have been retained from the original parent population, define that species.

Hallmark traits are both physical and behavioural. They are born out of the unique ecological pressures that the species has faced in its journey through its own natural history to become what is presently defining it. By their very definition, hallmark traits are critical to the survival of a species. Without these traits a species wouldn't be able to survive in the face of these unique pressures. For example, a giraffe has a long neck, likely due to facing an environment whereby they could outcompete other species by browsing in high branches. This hallmark trait is both a defining feature of a giraffe and a trait that allows a giraffe to realise its niche. A giraffe born with a short neck would not be able to thrive in its niche and may very well not survive its environment.

Along with being gifted with these traits, individuals of a species must have the ability to utilise these hallmark traits. In the wild, an individual of a species who does not have the ability to utilise their hallmark traits is less likely to survive, let alone thrive. In 1966, at Gombe Stream in what is now Tanzania, the population of chimpanzees being studied by Jane Goodall, were victims of a polio outbreak. Some of the chimpanzees that survived the disease were left crippled. Unable to ascend the treetops in the forest, and thus, unable to realise an important component of their niche, they eventually died.

In artificial environments, even though an individual may not face the same ecological pressures that they would in the wild, the ability to utilise their hallmark traits is just as critical to their well-being and their ability to thrive. Evolution has determined an individual's natural inclination to utilise these traits. If this ability is hampered, the individual may face significant physical or psychological stress.

Population-Level Pressures

In addition to pressures that affect an entire species and lead to species-level adaptations and traits, populations within a species may face unique ecological pressures. For example, a difficult-to-obtain preferred food source may be present in only one environment that a species lives in. The population that lives in this environment may develop a population-level adaptation in order to reach this food source. There also may be predators present in just one environment. The population living in this environment may need to develop an adaptation that allows them to stay safe.

If the population pressure is significant enough, and the adaptations are significant enough, evolutionary changes may lead to **subspeciation** (defining a new subspecies out of a parent population), or even eventual speciation. In these cases we see physical adaptations developing. Take, for example, your own blood type. The fact that, as humans, we have different blood types was caused at some point in our natural history when we faced population-level pressures that allowed different blood types to be adaptive in different areas. However, these adaptations may not just be physical. In many cases these adaptations are behavioural. Some species of monkeys have alarm calls that are unique to populations that denote the presence of specific predators within their environments (thus allowing group mates to flee danger based on which predator is present).[1]

Population-level adaptations can come in two forms: They can be genetic behaviours, limited to that population, that get passed along like any other inherited trait. They can also be cultural, meaning animals are learning how to adapt to a particular ecological pressure from each other. The presence of **culture** has long been studied in mammals such as primates and cetaceans. However, scientists are also studying

[1] See Price et al., 2014 and Price et al., 2015.

culture in other classes of animals, such as bird species, fish species, and even reptile and amphibian species. Additionally, they study how critical the presence of culture is to the survival of that species. The ability of species to learn from each other in order to survive is foundationally important to survival in the face of population-level pressures.

As with speciation-level adaptations, animals in artificial environments must be able to utilise their population-level adaptations. Interestingly enough, these animals develop novel population-level adaptations to their artificial environment. This is especially true of cultural traits. Animals in captivity will exhibit behaviours, not just to captivity as a whole, but to their specific artificial environment. Animals may develop unique calls, have unique methods of play, or even utilise tools specific to their environment. All of this belies the importance of a proper social environment.

2.3   The Evolution of Needs

We can define **needs** as what one requires in order to both sustain and utilise one's physical and behavioural adaptations. These needs vary, not only from species to species, but also between individuals of the same species. A species evolve needs over the course of their natural history. As we have seen, adaptations evolve when a species encounters an ecological pressure they must mitigate in order to survive in a particular niche. A species evolves needs based on these adaptations (Figure 2.2).

A cheetah adapted to the ecological pressure of having very quick prey by evolving the physical ability to run very fast, and the behaviour to utilise this ability. As a result a cheetah needs to sustain this adaptation with physical energy and healthy enough legs and muscles. Additionally, a cheetah needs the ability to utilise this adaptation with an open environment that allows the space needed to achieve such speed.

At the beginning of their natural history, primates adapted to an arboreal lifestyle in the trees with forward facing eyes, thus giving them extreme depth perception in the form of stereoscopic vision. Even though many primates eventually became terrestrial, they still retain this adaptation and utilise this adaptation for other critical means. In fact, stereoscopic vision has led to other adaptations. Without stereoscopic

Figure 2.2 Leopard at the Chattanooga Zoo. (Photo credit: author).

vision, many of the hallmark traits of almost all primate species could not be fully realised. Therefore, primates have evolved a need to both sustain and utilise stereoscopic vision. In order to sustain this trait, primates need the function of both eyes. In order to utilise this adaptation, primates need to live in an environment where they can benefit from the depth perception that stereoscopic vision allows.

Needs, like the adaptations they support, take the form of both physical and behavioural needs. Physical needs are most often those that allow adaptations to be sustained. While behavioural needs are most often those that allow adaptations to be utilised. However, this is not always the case.

For animals under human care, physical needs can be met by a sound nutrition plan, veterinary care and examinations, the proper sanitation of an environment, the proper physical structure of an environment, or anything else that bolsters the life functions involved in a specific adaptation. Sometimes physical needs of some species may be difficult, if not impossible, to maintain in artificial environments. In all cases, physical needs must be met for an animal under human care to have proper welfare.

Maintaining behavioural needs can oftentimes be more complicated than providing for physical needs. This may involve providing a proper social environment, behavioural welfare assessments, stimulating enrichments, and (as with physical needs) a proper environment that suits the specific species. Again, it may be difficult, if not impossible, to provide for some species' behavioural needs in a captive environment. However, it is critical for an animal under human care to have their behavioural needs provided for and maintained at all times.

Facultative Needs

At the species level, needs that are steeped in natural history are fixed throughout the course of one's lifetime. However, at the individual level, needs oftentimes are facultative; meaning that they can change within an individual's lifetime based on an individual's set of circumstances. **Facultative needs** are dictated by both the internal and the external, resulting in physical and behavioural needs. For example, an injury may cause an individual to require a certain physical need in order to fully realise their adaptations. A history of trauma may cause an individual to require a behavioural need.

Facultative needs are always critical to the general well-being of an individual. In some cases they may be essential to the very survival of the individual. However, even in cases where one might survive without their facultative needs, they are still essential to their general welfare.

In order to ascertain an individual's facultative needs, one must possess a comprehensive knowledge of that individual's history – medical, social, psychological, etc. Knowledge of an individual's history of injuries, periods of psychological trauma, any sensory deprivation, any social deprivation, environmental history, and anything else that may be relevant is critical in understanding an individual's current state of needs.

If these histories are not available, it is the responsibility of the care technician to perform assessments to ascertain what these facultative needs may be. First and foremost, this includes physical examinations to reveal past and current physical circumstances. This also includes behavioural assessments to determine what challenges the individual is facing in order to realise their adaptations.

It is critical to understand that facultative needs can be ever changing. They can also change very rapidly based on circumstances and internal and external factors. Facultative needs can change positively or negatively. Therefore, all care programmes must include methods of physical and behavioural assessments to determine what these facultative needs may be.

2.4   Basic Needs versus Thriving

Basic needs represent the minimum resources one needs to survive. In a strictly biological sense, basic needs include any resource that allows an individual to eat, stay safe, and stay healthy. At the population level, basic needs also include the ability to reproduce. Without access to basic needs, an individual will not survive, and a population will go extinct. However, it is critical to not confuse basic needs with what is required for thriving. It is entirely possible for an individual to have access to all of their basic needs and live in a state of poor well-being.

It is critical to distinguish between the importance of providing for an animal's basic needs versus providing an animal with opportunities to thrive. While basic needs allow an animal to survive, thriving allows an animal to live life to its fullest potential. Both must be provided to responsibly and ethically provide care to any living individual. Though different, basic needs and thriving are intrinsically linked. This is due to the fact that, not only are basic needs a requirement for survival, and thus thriving, but also that basic needs have largely determined the adaptations that are at the foundations of thriving. Providing basic needs becomes the first step towards allowing an individual to thrive.

The acquisition of basic needs is often seen as the prime mover of biology. The pursuit of food, the pursuit of safety, the pursuit of health, and the pursuit of reproduction, are all drivers of the evolutionary forces that create biological adaptations. As such, they represent not just the necessities for survival, but also the seeds of what is necessary for thriving. Most of the opportunities for thriving have their roots in the pursuit of those basic needs. The pursuit of the answers to: *What helps me to not be hungry? What helps me to stay safe? What helps me to stay healthy? What helps me to reproduce?* are the factors that determine evolutionary traits – both physical and behavioural. As such, they determine what an animal requires to thrive.

Categories of Basic Needs and Their Relationship to Aspects of Thriving

For any biological entity, basic needs fall into four categories: the need for food and water; the need to stay safe; the need to stay healthy; and (at the population level) the need to reproduce.

- *Satiance*

  Satiance is the feeling of well-being due to an absence of hunger. The feeling of satiance denotes that one is properly fed with enough calories to produce the proper amount of energy it takes to sustain life functions. Satiance itself is a basic need. From this basic need we find that a species, over the course of its natural history, evolves adaptations by choosing which foods are preferred, finding methods of obtaining food, and finding methods of defending that food. This has, in turn, determined many of the adaptations that are the hallmark traits of the species.

  While satiance fulfils a basic need, for thriving, an animal needs more. An animal requires a diet that allows them to utilise their adaptations in obtaining this food. Thus, for a diet to be prepared that promotes thriving, it must be thoughtfully presented, with the same variety the species might find in the wild, and with methods of obtaining and defending that food source that would mirror what the animal would find in the wild.

- *Staying safe*

  Freedom from grave dangers and the proper shelter from environmental elements are a basic need. Species have evolved methods of predator defence, methods of moving throughout their environment, and methods of seeking shelter that have allowed the species to survive and have created more of the species' hallmark traits. To fulfil basic needs in an artificial environment, proper shelter must be given, a safe environment must be provided, and an environment free of natural predators must be provided. However, in order to provide for an individual to thrive, animals must be given a naturalistic environment with places to hide, places to locomote naturally, and with the same elements of safe haven that a species would find in the wild.

  It is also important for animals to be able to observe their natural circadian rhythms and be active at the times of day that they are normally active. Such behaviour usually has its roots in either staying safe from predators or in finding prey and food sources – both of which are necessary components of basic needs.

- *Staying healthy*

  While staying healthy through a proper diet, proper veterinary care, and proper exercise is a basic need, *feeling healthy* is an important measure of thriving. Feeling healthy would include being able to move about in a way that an animal is naturally inclined to move about. It would also include having access to the outdoors that an animal would encounter in the wild. It would include an environment that is free of foreign odours, noise pollution, light pollution, or the absence of enough light. Feeling healthy is an essential component of welfare and a critical factor of thriving.

- *Reproducing*

  The ability to reproduce is a basic need only at the species or population level. An individual can survive just fine without ever reproducing. Obviously, however, if a population doesn't reproduce, it goes extinct and is, therefore, a basic need. However, the ability for natural reproductive behaviours is an essential component of thriving for animals in artificial environments. As such, the ability to interact and socialise with conspecifics is foundational to this ability. Also, providing the freedom of choice to choose which conspecifics one interacts with becomes a necessary building block for thriving.

2.5   Natural Ecology

In captivity, American robins have been shown to alter their food preferences based on seasons.[2] Captive coyotes show a significant increase in species-typical behaviours when feeding times and presentations are made to be unpredictable.[3] The gut microbiome (the ecosystem of microorganisms that live in the digestive tract and aid in digestion) of lemurs are shown to vary significantly between wild habitats and captive habitats.[4] Fish that are born in captivity then released into the wild face significant disadvantages in terms of survival unless they are hatched in a captive environment that mirrors a natural setting.[5] What these four examples have in common is that they all relate to the natural ecology of a species and how captivity can severely affect an animal's well-being based on how well an artificial environment mirrors that ecology.

An individual's **natural ecology** is defined by their interactions with the natural environment in which they live, both internally and externally. What an individual consumes, excretes, takes from an environment, or leaves behind are all facets of one's natural ecology. The presence of each organism in an environment, along with every action that an organism performs, the structure of an environment, and what can be perceived by the senses in an environment, dictates a natural ecology.

Perhaps the most significant challenge of captivity is that it is impossible to reproduce all aspects of one's natural ecology in a captive environment. Because captivity removes an animal from a natural ecology, their adaptations, forged throughout their natural history, can be made either less significant or inadequate. This can greatly alter an individual's welfare. No captive environment can provide all of the components that a species would face in the wild. However, captivity *can* provide factors in a species' natural ecology that are the most significant and attempt to reproduce them within the captive environment. In order to do this, a care technician must possess a knowledge of the internal and external elements of specific natural ecologies.

Natural ecology can be thought of as a ship on the water. Different ships are built for different circumstances. Some may be built for the high seas. Some may be built for shallow water. Some may be built to break through ice. Others may be built to withstand the tropics. Some ships are built for salt water. Other ships are built for freshwater. Ships depend on their internal mechanisms, such as the shape in which

---

[2] See Wheelwright, 1988.
[3] See Gilbert-Norton et al., 2009.
[4] See Bornbusch et al., 2022.
[5] See Johnsson et al., 2014.

they are constructed, the materials they are constructed with, and the sealant on those materials so they don't take in water. Their construction determines what they can hold and how much they can hold. If a freshwater boat attempted to travel in the ocean, it might corrode and sink. If a light boat attempted to hold heavy cargo, it would sink.

Ships are also dependent on their external mechanisms. Some ships have sails that dictate how they behave in certain winds. Some ships have large paddles that can propel them through a certain type of water. Ships must be steered by a reliable captain. Ships must have the ability to deal with, or steer away from, unforeseen circumstances that may arise on their journey. A paddle boat would not be able to behave in a way that would allow it to survive in the windy high seas. A sailboat without a savvy mariner, would not be able to utilise the wind effectively. As with the internal features of a ship, the external determines what type of water it can survive.

All biological entities are endowed with the same internal and external mechanisms to navigate their own natural ecologies.

Internal Mechanisms of Natural Ecology

Internal mechanisms of natural ecology relate to how an organism has physically adapted to interact with their specific environment. For example, an animal may have a digestive system that is suited to a particular type of diet that is dictated by the food that is found in their unique environment. Add to this the gut microbiome of each organism – where specific microorganisms reside in the digestive tract in order to process the very specific food items found in this environment.

Internal mechanisms also determine the natural energy budgets that are needed to survive an environment. An animal that needs to traverse through more difficult terrain may require a higher energy budget than an animal that has an easier terrain or smaller area. This can be granted to them by how quickly they can convert food into energy and, of course, what food sources they prefer.

Internal mechanisms can dictate what physical features an animal may have to obtain food resources. A strong, beaked bird has the ability to crush through food with hard shells. A bird with a long thin beak has the ability to obtain food sources found within tiny crevices.

Even the physical structure of different individuals can be due to internal mechanisms of natural ecology. The amount of body fat, the length of extremities, and the general build of an individual's frame can all relate to how an animal must be constructed to survive in different environments.

Internal mechanisms of natural ecology are essential to an individual's survival. Possessing these mechanisms and being able to use them, allow an individual to live safely and healthily. When that environment changes, these internal mechanisms may no longer be relevant; and therefore an animal may not survive. It is therefore critically important that when an animal is placed in an artificial environment, that every measure of that environment must be created to allow for the relevance of those internal mechanisms.

External Mechanisms of Natural Ecology

How an individual actively interacts with the environment around them and behaviourally adapts to the circumstances present in that environment, along with navigating through new and unforeseen circumstances, represents an individual's external mechanisms. How an animal responds to predators in an area, where to find food and water, where to seek shelter, and how they interact within their population with regard to gaining protection from conspecifics and reproducing, determines the survivability of an individual and population in their specific environment.

External mechanisms can be both fixed and plastic (having the ability to adjust based on experiences). External mechanisms can be both fixed or cultural (passed between conspecifics through social transmission). Most external mechanisms, however, are deeply ingrained within a species. All living organisms have instinctive actions that are based on their ecology. As with external mechanisms, when animals are placed in artificial environments, these internal mechanisms may no longer be relevant. They can even prove to be maladaptive to the individual. It is therefore critical for environments to be constructed in such a way that external mechanisms of natural ecology stay relevant for the general well-being and survivability of the individual.

Implications

What an individual requires to thrive is seeded in the very natural history of the species. It is a chain that has been forged link by link by the forces of evolution and adapting to specific ecologies. It rests in the basic needs of the species and exists in the facultative needs of the population and the individual. Some aspects of thriving are fixed within a lifetime of an individual. However, others change based on the environment and ecology.

In the wild, animals are endowed with a natural process of finding opportunities to thrive based on a natural ecology. However, when an animal is removed from nature, by either capture or captive birth, and placed in an artificial environment, they lose their unique, species-specific, ability to interact with the natural world around them due to the fact that they are not in the environment where this ability evolved. Therefore, natural ecologies must be replicated in artificial environments so that an animal has the opportunity to thrive and is able to utilise their adaptations in a manner that promotes the general well-being and welfare of the individual.

A true care technician has a comprehensive knowledge of the natural history of the species, of the natural ecology of the species, and a working knowledge of the individual – both medically and behaviourally, and the ability to assess their current circumstances. Then, and only then, can true welfare be provided and obtained.

References

Bornbusch, S. L., Greene, L. K., Rahobilalaina, S., Calkins, S., Rothman, R. S., Clarke, T. A., LaFleur, M. and Drea, C. M., 2022. Gut microbiota of ring-tailed lemurs (Lemur catta) vary across natural and captive populations and correlate with environmental microbiota. *Animal Microbiome*, 4(1), pp. 1–19.

Gilbert-Norton, L. B., Leaver, L. A. and Shivik, J. A., 2009. The effect of randomly altering the time and location of feeding on the behaviour of captive coyotes (Canis latrans). *Applied Animal Behaviour Science*, 120(3–4), pp. 179–185.

Johnsson, J. I., Brockmark, S. and Näslund, J., 2014. Environmental effects on behavioural development consequences for fitness of captive-reared fishes in the wild. *Journal of Fish Biology*, 85(6), pp. 1946–1971.

Price, T., Ndiaye, O., Hammerschmidt, K. and Fischer, J., 2014. Limited geographic variation in the acoustic structure of and responses to adult male alarm barks of African green monkeys. *Behavioural Ecology and Sociobiology*, 68(5), pp. 815–825.

Price, T., Wadewitz, P., Cheney, D., Seyfarth, R., Hammerschmidt, K. and Fischer, J., 2015. Vervets revisited: a quantitative analysis of alarm call structure and context specificity. *Scientific Reports*, 5(1), pp. 1–11.

Wheelwright, N. T., 1988. Seasonal changes in food preferences of American Robins in captivity. *The Auk*, 105(2), pp. 374–378.

3   Measures of Thriving

Figure 3.1 The author meets meerkats at the Suffolk Owl Sanctuary. (Photo credit: author's collection).

3.1   A Lighthouse Keeper's Tale

Through the dense fog and rain of a stormy night, a sailor spots a beacon of light on a distant shore. He shifts his rudder and adjusts his sail accordingly. Putting his vessel on a path to reach, what he assumes, is a safe docking area and shelter from the storm. As he draws nearer, the light seems to shift. He no longer appears to be on the right course. Quickly, he scrambles, trying in vain to adjust his navigation to get back on course. However, the light shifts again, distorted by the fog. The wind picks up again, rocking the boat wildly. Suddenly, the sailor is thrown from his ship, as it crashes against an outcropping of rocks. Near death and fading from consciousness, the sailor looks up to see that the beacon of light was indeed coming from a lighthouse. As darkness overtakes him, his remaining thoughts are of how the lighthouse could have led him so astray.

The sailor does not die. In fact, his recovery only takes a few weeks. When he is back to full health, he vows that no other sailor should ever have to experience his fate. He begins to make plans to purchase the lighthouse so that he can fix whatever is wrong with it.

Once he acquires the lighthouse, he realises that he needs advice. He stops by a local tavern, frequented by the lighthouse keepers along the coast. He relays his harrowing tale. One of the lighthouse keepers says to him that his beam of light is likely to dim. He should increase the gas, which will increase the flame. He should also clean the glass panes that house the lame.

Excitedly, the sailor rushes back to the lighthouse. He first examines the flame. He doesn't know how bright a lighthouse flame should be, but he trusts that the lighthouse keepers know what to do. As such, he increases the gas. Sure enough, the flame gets much brighter. Next, he cleans the glass panes. When he is done, he is impressed with how brightly the light shines across the water and across the rocky shore. Feeling quite satisfied with himself, he goes downstairs for a bowl of chowder. There, he will await the fall of night.

Unfortunately, the day's hard work and the warmth of the chowder make him sleepy. He drifts off to sleep and does not wake until the light of morning. When he goes outside, he is horrified to discover that three ships have crashed along the banks. He rushes to see if there are any survivors, or any hints of why the lighthouse didn't work. However, all he finds are the wrecked debris of ships and the mangled bodies of dead sailors. Overcome with grief and guilt, the sailor returns to the lighthouse.

Consumed by his failure, the sailor returns to the tavern. A half bottle of gin later, he makes his confession to everyone in the tavern.

'It is I who is solely responsible for the deaths of the sailors last night', he cries. 'I took on the lighthouse without the knowledge of how one maintains a lighthouse or how one makes a lighthouse functional'.

Out of a dark corner of the lighthouse, he hears an old man scoff. Out of the shadows, a bearded old salt appears.

'I am the oldest lighthouse keeper in this tavern', said the old man. 'I've been tending the flame since before any of these other men were born. I've seen the worst of shipwrecks. I've seen what works. I have seen what fails'.

The sailor pleads with him to help. The lighthouse keeper agrees on one condition: he must follow his every word. The lives of sailors depend on it.

When they return to the lighthouse, the lighthouse keeper asks him exactly what happened. The sailor explained that he made the light brighter by increasing the gas. He cleaned the glass panes. Even with all of these efforts, the lighthouse failed. In fact, said the sailor, things got worse.

'You made assumptions about the lighthouse that weren't the case', said the lighthouse keeper. 'You cannot tell what a lighthouse needs without first observing it, collecting evidence, and then drawing a conclusion based on that evidence of how to fix what is wrong'.

The sailor asked what he should do.

The lighthouse keeper respond that he should light the lamp, go out into the water, and note what he observes.

The sailor takes a small schooner into the water and observes the lighthouse from different angles. With the benefit of the light of day, perhaps he can gain a different perspective of what is occurring. However, nothing seems wrong. He starts to bring the schooner back to shore.

The lighthouse keeper shouts, 'No! Go back! You need to see the lighthouse beacon in all manner of light!'

As the sun begins to fall in the sky, and the blanket of darkness begins to make the lighthouse shine, the sailor notices something peculiar. The beacon of light, though bright, is acting strangely. When the light hits a large outcropping of rocks on the shore, it reflects. He surmises that if it gets foggy enough, the reflection would create a second beacon of light. This, in turn, would lead sailors directly into the rocks.

Excitedly, he returns back to the lighthouse to tell the lighthouse keeper what he has found. The lighthouse keeper instructs him to remove the rocks from the shore. Into the small hours of the night, the sailor works tirelessly to gather and move the rocks to a place where they will no longer reflect the light.

Exhausted, he returns to the lighthouse keeper and exclaims that they can finally light the beacon. The lighthouse will finally be in order.

'We are not nearly through', says the lighthouse keeper. 'Have you actually looked at the components of your lighthouse? Have you actually looked at what is inside the machinery?'

The sailor admitted that he had not.

Under the lighthouse keeper's direction, he looked at the inner workings of the lighthouse. He showed the lighthouse keeper each bit of machinery. Each time, the lighthouse keeper would look at it, and tell him it was in working order. However, when he looked at the actual light, the lighthouse keeper stopped him.

The light, he says, is actually too bright. It is far brighter than any lighthouse he has ever seen. The flame is too hot.

'By increasing the heat in the light, you made it brighter', says the lighthouse keeper. 'You made the beacon of light reflect more than it had before. This was one of the reasons why so many sailors died.'

Together, they repair what the sailor had done to the lighthouse the previous day.

'So now', says the sailor, 'we have removed the shiny rocks and we have restored the light to its proper brightness. Can we finally light the lamp and steer the sailors to safety'.

'No', says the lighthouse keeper. 'You need to find out the history of the lighthouse and see if there is any other information that is important. This is critical'.

'But how', asks the sailor.

'Let's find the logs from the old keepers', replies the lighthouse keeper.

Together they search throughout the lighthouse until finally they come upon several dusty volumes of lighthouse logs. As they pour through the logs, they begin to notice a pattern. The logs note that on different days of the month, the lighthouse keepers must block a side of the beacon. As they read further into the logs, they realise that, due to the light of the moon, the lighthouse beacon can reflect off different portions of the shore. The manner in which to mitigate this is to shield a portion of the beacon from where it will reflect on a particular night.

Smiling, the lighthouse keeper closes the logs. 'Now', he says, 'you have the tools you need to create a plan for how to operate your lighthouse. I, myself, am going back to the tavern'.

Assessing Animal Welfare

In our story, the sailor's quest for redemption begins with a crisis. This crisis centres on the fact that he knows something must be done to make the lighthouse function properly. However, he doesn't understand what is wrong with the lighthouse.

Nor does he understand how to properly respond to lighthouse issues even if he did know what was wrong. Very quickly he learns two critical points:

- Each lighthouse is unique with unique issues.
- Without a proper knowledge of what is uniquely wrong with a specific lighthouse, he can make matters much worse – sometimes *critically* worse.

The sailor's quest to fix his lighthouse mirrors anyone who endeavours to create real-world solutions to solve real-world issues. His story certainly mirrors a care technician who is tasked with the care and well-being of individual animals in a collection. As care technicians, we must start by acknowledging that each animal is an individual, with individual needs, individual challenges, and individual histories. Like the sailor, as care technicians, we must mitigate those challenges and meet those needs through evidence. Like the sailor in our story, we can find this evidence through a scientific process of observation and assessment. Finally, like the sailor, we also must gather what we can deem **a priori**, or pre-observational facts, about the individuals in our charge. This combination of observational assessment and a priori knowledge of individuals is the key to evidence-based individualised care, which is foundational to allowing animals to thrive.

Observational measures and assessments are critical means by which a care technician can gain an understanding of the current state of the individual as well as how both external and internal factors are affecting that individual. Because individual challenges and needs are fluid, and can change significantly over a short period of time, observational measures and assessments need to be performed regularly and often. The scope of observational measures and assessments ranges from simple anecdotal observations to complete physical examinations. However, all aspects of this spectrum should count as evidence and should, therefore, proceed according to a scientific method. For the purposes of this text, we will divide these measures into behavioural observations and physical observations. Though complete veterinary examinations are also part of this means, the techniques therein are outside of the scope of this book and rest in the domain of veterinary science.

Baselines

Because each individual is unique, with a unique history, unique challenges, and unique needs, it is important to establish individual baselines. Baselines exist as a set of data that shows what is typical for that individual. Baselines are both physical and behavioural. They are established through repeated assessment periods and physical examinations. Once an individual's baseline has been established, a care technician can see positive or negative changes in this baseline during future assessments. This is keenly important when looking at the effectiveness of new care plans, new enrichment items, a new medical script, an ailment an individual may be suffering from, or any other external or internal factors that may be affecting an individual's general well-being. It is important to note that just because something is baseline for an individual, doesn't mean that it is satisfactory for an individual. It is merely a tool to establish what behavioural or physical condition can be expected from that individual.[1]

It is important to take a mathematical approach when establishing baselines. For physical baselines, this may just involve determining what a typical blood pressure, weight, EKG, and so on is for an individual. This becomes more complex when looking at behavioural baselines. Establishing behavioural baselines requires determining expected frequencies of behaviours over a period of time or determining which behaviour is most likely to occur in the presence of a given stimuli. This can also include feeding behaviours such as caloric intake or food preferences.

Whatever method is employed to arrive at baselines, it must be consistent. Without a consistent methodology, a care technician is in danger of creating faulty or misleading baselines and may be missing critical information about that individual's well-being.

It is important to note that baselines are fluid, and may change with each assessment period. The more assessment, the more information there is to compile the baseline. Additionally, behaviours and physical states change when internal and external factors change. Therefore this fluidity is expected as animals adjust.

3.2   Behavioural Assessments

The manner in which animals interact with the world around them, and react to any given stimuli, is collectively prime indicators of general well-being. Oftentimes before an individual shows any physical signs of physical changes to their welfare, they're behaving in a way that is indicative of this change in welfare. A scientific approach to assessing behaviour is the cornerstone of measures of thriving (Miller et al. 2020). Behavioural assessments are non-invasive, cause little to no stress to the animal or population being observed, are simple to perform, and can be done with great regularity. In some cases behavioural observations involve **ad libitum** note taking (narrative descriptions) regarding daily observations. Other types of behavioural assessments may be more involved; utilising ethology, behavioural indicators, or complex social and spatial charts. However, no matter how simple or complex, it is critically important that behavioural observations take a scientific approach.

Observer Reliability

Unlike many physical assessments, behavioural observations can be fraught with subjectivity and observer bias. Observer bias occurs when the observer favours placing either too much significance or not enough significance on a behavioural episode as compared with other observers. Such cases lead to inconsistencies in reporting and can greatly skew results of any assessment.

---

[1] See Hill & Broom, 2009.

A solution to this is known as reliability tests. Reliability is established when a team of observers can watch the same behavioural episode and record the same data point. An effective way to perform a reliability test is to record a video of different behavioural episodes and ask all observers on a team to score the video. Reliability can only be established at a 95 per cent similarity rate. It is critical that such a test be performed before assessments can be deemed accurate.

Anecdotal Observations

Anecdotal observations of animal behaviour are simple accounts of occurrences (Figure 3.2). They are often the first indicator of anything unusual within an individual or population. Though these observations are simple, they can and should proceed just as scientifically as any other observational method and should be included as part of any behavioural assessment programme.

Though, on the face of it, anecdotal observations are not immediately quantifiable, the evidence they provide can be culled in such a manner as to discern patterns of behaviour over a period of time. Because of this, there needs to be a systematic approach to recording these observations. Daily log keeping for all care technicians is a must. Within these daily logs, notes and observations should be recorded every day. These should include ad libitum descriptive accounts of an animal's behaviour every day, no matter how seemingly insignificant. Over time, these notes and observations can reveal patterns that may precede changes in an individual's behavioural baselines. Once recorded, these observations should be stored permanently and be able to be quickly searched and referenced when needed. Paper logs should be filed and catalogued in a manner that is easy to access. Digital files must be backed up and exist in a software platform that allows for ID searches, keyword searches, and category searches.

Oftentimes we see documentation of new and novel behaviours first appearing in these anecdotal observations. Sometimes we may see a disturbing change in an individual's behaviour. These can then be used to trigger more complex assessments where a complete picture of an individual's situation can be garnered.

Formal Care and Welfare Assessments

Formal care and welfare assessments are a process by which observable behavioural indicators are scored each day by care technicians over a set period of time. After the assessment period, the results are quantified, analysed, and conclusions are drawn as to the state of well-being of an individual animal. This is also used to compare the current state of an individual to previous baselines. Behavioural indicators are species-specific; determined by the typical behaviours of the species, such as typical range of motion, activity budget, feeding behaviour, and socialisation. However, these indicators measure categories of well-being which do not vary across species. Rather, they are the uniform tenets of positive well-being.

Figure 3.2 Students observe chimpanzees at Lion Country Safari. (Photo credit: author).

The process of these formal assessments goes like this: an assessment period is defined (usually something like a 60-day period, however, this can vary based on species or circumstance). Care technicians are given a list of indicators that they are to watch for during a set time. This set time may be an entire shift, or it may be preselected times of the day. At the end of this time, the care technician scores these indicators in a preselected manner (usually utilising a 5-point numeric scale, see below). After the assessment period, all the data is quantified, statistically analysed, and conclusions are drawn based on these results. These conclusions are used to determine the effectiveness of current care plans, determine if new care plans are required, or if there are signs of a condition that requires more serious intervention.

Each individual has an assessment on a rolling basis. They may occur semi-annually, annually, bi-annually, or by whichever timeline is established by the care protocols.

- *Categories of indicators*

    Though the specific indicators are going to vary from species to species; and perhaps, vary from population to population (based specifics such as typical socialisation, typical feeding behaviours, and typical activity budget), the categories of what one is measuring with these indicators is uniform across species. These categories are: mental domain, feeding behaviours, movement, evidence of physical distress, environmental interactions, socialisation, and stereotypic behaviours. They can be defined by the following:

    o Mental domain: Is the individual behaving in such a way that shows mental acuity; or, rather, does the individual appear disoriented, confused, or perform behaviours that reveal a lack of mental function at given stimuli?

    o Feeding behaviours: Is the individual consuming a typical diet? Is the individual anxious when food is present? Is the individual showing a normal food preference?

    o Movement: Is the individual performing a range of movement that is typical for that species? Is the individual utilising a typical activity budget?

    o Behavioural evidence of physical distress: Is the individual behaving in such a way that would denote physical distress?

    o Environmental interactions: Is the individual interacting in a species-typical manner with their habitat? Enrichment? Is the individual making use of a typically preferred space? Is the individual exploring their environment in a species-typical manner?

    o Socialisation: Is the individual socialising with conspecifics in a species-typical manner? Is there elevated aggression? Is there anxiety associated with socialisation?

    o Stereotypic behaviours: Is the individual engaged in behaviour that is repetitive without an immediately identifiable goal? Is the individual involved in self-injurious behaviour?

    Within each category, specific indicators will properly measure what is typical behaviour for the studied species. It is incumbent on the care technician, when arriving at these indicators, to properly research the species and to adequately cite this research within the descriptors of each indicator. This will ensure that the proper indicators are being used and are effectively measuring the behaviours within each category.

- *Scoring*

    The exact method one employs to score behavioural indicators can vary based on the information one is attempting to ascertain. However, in most cases, indicators are scored by numeric scale. The scale ascends from behaviours indicative of poor well-being to behaviours indicative of superior well-being. In a 5-point numeric scale, a score of '1' would indicate negative welfare, while a score of '5' would indicate positive welfare. A score of '3' would indicate baseline behaviour. Five-point scales measure both negative changes in welfare and positive changes in welfare.

For example, if one is looking at an individual's amount of stereotypic behaviour using this 5-point scale, the individual's baseline amount would be scored a '3'. Let's assume the baseline is three episodes of stereotypic behaviour per day. Because a greater amount of stereotypic behaviours indicates a negative change in well-being, on days that the individual has any more than three observed episodes, would result in a score of '2' or '1' (if there is a predetermined significantly higher amount of these behaviours). On the other hand, on days where only 1 or 2 episodes are observed, the individual would receive a score of '4'. If no episodes are observed, the individual would receive a score of '5'. After the assessment period is concluded, this data is quantified to determine if there is a positive or negative change to the baseline behaviour of that particular indicator.

A 3-point scale can also be used. The advantage of using a 3-point scale rather than a 5-point scale is that with fewer choices, there is less chance of subjectivity when scoring these indicators. Thus, reliability increases. The disadvantage of a 3-point scale is that it only measures negative changes to a baseline. In these scales, a score of '3' still represents a baseline, while '2' represents a slight negative change, and '1' represents a significant negative change.

- *Quantifying results*

  Properly quantifying and assessing the raw data is, perhaps, the most critical step. Improperly quantifying the result can lead to a care technician misidentifying a welfare issue, missing a welfare issue, or determining an issue that may not exist. In order to properly assess results, a mathematical approach is required. Luckily, this can be accomplished with easy graphing and simple statistics. Additionally, once the framework is established, quantifying results is a fairly quick step.

  Let's imagine a scenario where a chimpanzee is undergoing a 60-day care and welfare assessment. We will use a 5-point scale, where 5 represents the most positive indication of welfare, 3 is baseline, and 1 represents the most negative indication of welfare. For the purposes of this example, let's keep these indicators simple and only utilise four categories: environmental interactions, socialisation, mental domain, and movement. We will create simple indicators to serve these categories (Box 3.1).

  After 60 days, each indicator has a certain amount of 1's, 2's, 3's, 4's, and 5's. For each category, we will average those together to come up with a mean for each category. For example, for the 'Environment' category, let's say that, over the 60-day period, the amount of days that this chimpanzee scored a '3' (baseline) for 'Interaction with Environment', 'Engagement with Environment', and 'Use of Available Space', averages to 32 days. Therefore, the category score for Environment is 32. Once these numbers are derived, a table can be created (Table 3.1).

  Once we create our table, we can create a visual representation of the indicators. Because a score of 3 represents the individual's baseline. We would expect a distribution that looks like a bell curve.

Box 3.1 Sample categories and indicators
Care and Welfare Indicators and Categories

- Environment
  - Interaction with environment
  - Engagement with enrichment
  - Use of available space
- Social behaviours
  - Social proximity
  - Agonistic behaviours
  - Affiliative behaviours (e.g., Grooming)
  - Courtship and reproductive behaviours
- Mental domain
  - Interest
  - Sensory
  - General awareness
- Movement
  - Activity level
  - Use of extremities
  - Pattern of sleep

Table 3.1 Sample table for categorical scores

Scale	Environment	Social	Mental domain	Movement	Total
1	4	1	20	0	25
2	23	6	26	0	55
3	32	46	10	52	140
4	1	6	4	4	15
5	0	1	0	4	5

In Figure 3.3, we see the overall total scores for this individual, form a normal bell distribution, with the baseline score dominating the other scores. Overall, this distribution represents an individual operating according to baseline behaviours. However, a deeper look at the total distribution reveals that scores of negative welfare outnumber scores of positive welfare. If we look even closer, we see that this skew is largely due to the mental domain category, where the 2 scores are relatively high. Perhaps this is an older individual, in relatively good physical health and socially active, but in mental decline. This could help to inform future care plans.

To further explore the significance of these scores, we may choose to employ a statistical formula such as a **goodness of fit test**. Running such a test on the overall total scores of the indicator categories reveals that the distribution is normal and not significantly scored to the left (negative welfare) or to the right (positive welfare). So we can infer, with statistical confidence, that this individual is, overall,

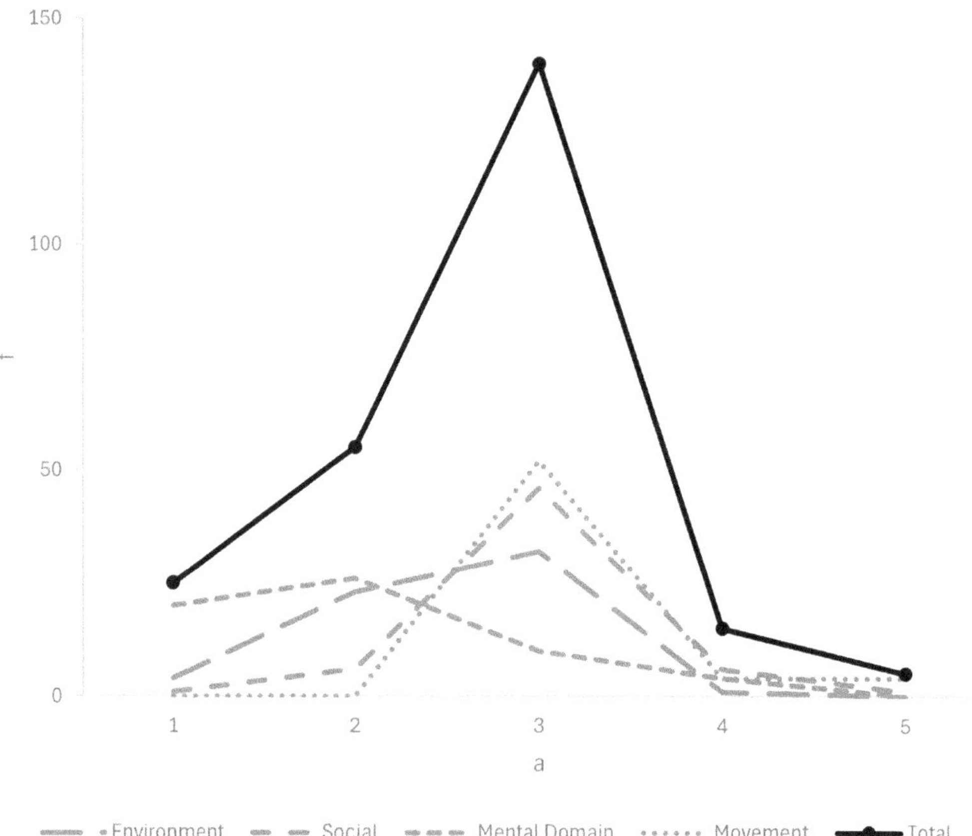

Figure 3.3 Visual representation of indicator category scores. The solid line represents the total overall welfare score, the dotted lines underneath represent environment, social, mental domain, and movement, respectively.

exhibiting baseline welfare. However, if we look at the categories individually, we may see a significant skew. In our example, the largest skew is mental domain. However, running a goodness of fit test on this sample still reveals no statistically significant skew. However, even without statistical significance, we should not assume that there is no cause for alarm. Rather, it is advisable to continue to monitor this individual and collect more data, especially with regard to mental domain. At the same time, we should not be stopped from acting on the apparent (albeit not statistically significant) skew towards negative welfare. Instead, this evidence should inform future care plans.

- *Drawing conclusions from results*

    Quantified results of behavioural indicators only give a dry, context-free, snapshot of what is occurring during that assessment period for that particular individual. This itself is not a conclusion. Making sense of this raw data means to look at the contextual history of that individual; drawing from internal and external factors that have been occurring during the assessment period and drawing on a knowledge of that individual's baseline and what has been traditionally known about that baseline. This knowledge, combined with the raw data, creates the conclusion that this behavioural assessment has drawn.

This conclusion can take the form of a report that is distributed to all care technicians and involved veterinarians. This report can then be used to make a case for either a new or amended care plan or a sustained care plan. In this way, the care of the animals, and the protocols therein, is based on evidence.

- *Ad hoc, or triggered, assessments*

    In some cases, formal care and welfare assessments may need to be performed outside of an individual's normal assessment time. They may be triggered by an external or internal factor which requires a deeper examination of an individual's welfare at that point in time. Triggered assessments should proceed exactly like the regular formal assessment; utilising the same process of gathering data from indicators and drawing conclusions based on that data.

    Assessments should be triggered based on predetermined parameters. For example, a triggered assessment may occur after an anecdotal observation of a concerning behaviour. An assessment may occur after an addition to loss to a population. These parameters should be defined by anything that might denote a change in welfare.

    Lastly, a triggered assessment should never replace a regular formal assessment. Rather, it should serve as an additional assessment.

Ethological Observations

When we discuss an animal's behaviour, in general, we are not speaking about one behaviour, or one instance of a behaviour. Rather, we are talking about a continuous series of behaviours that collectively make up our perceptions of how an animal is acting. The process of analysing how each behavioural instance creates an overall behavioural state of affairs is called **ethology**. Ethology can be an extremely useful tool in assessing the well-being of both an individual animal and the well-being of an entire population.[2]

The process for ethology consists of assembling a list of **event behaviours**, or instantaneous behaviours that occur as a short momentary occurrence (e.g., bark, jump, and yawn), and **state behaviours**, or behaviours that occur for a measurable amount of time (e.g., resting and socialising). This list should include all events and states that one may reasonably see in a given period of time relative to the species and circumstance. The list should also include both an event and state for 'other'. Together, this list is called an **ethogram**.

- *Ethological sampling*

    After an ethogram is derived, ethological observations can occur. Collecting this data is called sampling. Data is collected either by **focal sampling**, where the

---

[2] For a complete guide to effectively performing ethology, look no further than Robert Hinde's classic *Animal Behavior* book. See Hinde, 1966.

observer surveys one individual, or **scan sampling**, where the observer surveys a group of animals, and records which behaviour, listed in the ethogram, is being exhibited. Choosing between focal and scan sampling depends on the type of information desired.

- o *Focal sampling*
  - One individual is followed for a predetermined amount of time. During this time, only this individual's behaviour is being recorded. The behaviours of others may only be recorded if they are relevant to the actions of the focal individual (and if the data sheet allows for such recording). Behaviours that are not included in the ethogram may not be recorded. Depending on the type of information desired, individuals may be chosen at random or specifically.
- o *Scan sampling*
  - The behaviour of multiple individuals is recorded at once. In most cases, which individuals are being included in the scan sample must be determined before the observation. Only the individuals in this group may be recorded, unless the behaviours of outside individuals are relevant to the actions of the scanned group. Ethograms for scan sampling are usually very different from ethograms for focal sampling in that they oftentimes will include more proxemic data, or information relating to social or geographic spacing. Behaviours that are not included in the ethogram may not be recorded.

Once the type of sampling is decided, a sampling standard must be determined: opportunistic sampling or intermittent sampling. Both focal samples and scan samples can either be collected opportunistically or intermittently.

- o *Opportunistic sampling*
  - Taking opportunistic ethological data entails recording any observed behaviour listed on the ethogram when it occurs. Opportunistic data is best when one wants to know how many times an individual is exhibiting a certain behaviour as opposed to others. However, opportunistic sampling requires a much smaller ethogram and, therefore, should not be used to arrive at a comprehensive picture of an individual or group's behaviour.
- o *Intermittent sampling*
  - Taking intermittent ethological data entails choosing a recurring time period and recording whichever event and state behaviours, listed on the ethogram, are occurring at that exact moment. For example, if one is performing a one-minute focal sample on an individual, every minute, on the minute, the observer records which state and event is being exhibited by the individual at the moment the minute passes.
  - Intermittent sampling is the best method of arriving at a comprehensive picture of an individual or group's behaviour. However, it requires enough data to be collected in order to be properly quantified and analysed. For example, if one has an adequate sample size of focal samples for an individual, one can make evidence-based conclusions regarding that individual's general behavioural well-being. Similarly, intermittent scan samples can give a valuable assessment of a population's stability and social well-being.

- *Data sheets*

  Once the sampling methods and standards are decided upon, a method of data collection must be created. This involves creating a **data sheet**, or the method whereby the observer records ethological data. The way one composes a data sheet is a critical part of ethology. Generally, data sheets are in a spreadsheet format that includes a column for time, individual or group (if scan sampling is being employed), state behaviour, event behaviour, and notes. However, some questions that need to be answered before designing a data sheet are:
  - Which sampling method is being employed? If opportunistic sampling is being used, the data sheet might be structured differently than intermittent sampling. For example, if one may be most interested in the number of occurrences of a particular behaviour while doing opportunistic sampling, and therefore the data sheet may be more of a tally chart.
  - If intermittent sampling is being used, what is the interval between observations? This should be pre-recorded so that every interval is being observed.
  - Is focal sampling or scan sampling being employed? If scan sampling is being used, how are groups being identified on the data sheet? Is the data sheet only for one group?
  - How are states and events being abbreviated? Generally, behaviours are too long to write while effectively taking observational data. Therefore, an easily memorisable abbreviation system should be used.
  - Are notes desirable? One of the strengths of ethological data is that it removes much of the subjectivity of ad libitum data collection. Will adding ad libitum notes to a data sheet compromise the objectivity of the data?

  Once the structure of the data sheet is settled, the work of collecting data, the analysing and quantifying the results can begin (Figure 3.4).
- *Quantifying data in ethology*

  The body of data resulting from an ethological assessment can provide a wealth of information about both an individual and a population. One might be simply looking at a percentage of time an individual is engaged in certain behaviours. Conversely, one might be attempting to find out more specific information, such as measuring the significance of the actual time an individual is engaged in various behaviours versus what would be expected. As such, different quantification methods would be warranted.

  First, let's assume that the care technician is looking for a basic activity budget. For this, quantifying the results would require examining each behavioural state and determining the percentage of time each event occurred within that state during the observation period. For example, let's say that the care technician is observing an individual chimpanzee over a period of time. Breaking down all of the behavioural events in a 'social' behavioural state might look something like Table 3.2.

  Once these percentages are derived, conclusions can oftentimes be drawn just from a visual representation of this data. In the case of our focal chimpanzee, a visual representation might look like Figure 3.5.

Table 3.2 A sample breakdown of behavioural events within a social state

Event	Percentage
Groom (actor)	6
Groom (receiver)	2
Play (initiate)	3
Play (respond)	7
Communicate – distant (many channels)	4
Communicate – visual	13
Communicate – tactile (non-groom, non-aggression)	6
Communicate – chemical	1
Communicate – vocalise (non-play)	12
Initiate aggressions	2
Respond to aggression	4
Idle in proximity	8
Reassurance (ask)	9
Reassurance (give)	6
Flee	2
Display	6
Others	9

Time	ID	Location	State	Event	Notes
13:04	ELGIN	CTR CULVERT	REST / SONS	SLEEP	
13:05	ELGIN	CTR CULVERT	REST / SONS	SCRATCH	IRRITATED BY A WASP
13:06	ELGIN	CTR CULVERT	SIT / SONS	YAWN	
13:07	ELGIN	FRT LAWN	MOVE / SONS	KN WLK	
13:08					
13:09					
13:10					
13:11					
13:12					
13:13					
13:14					
13:15					
13:16					
13:17					
13:18					
13:19					
13:20					

Figure 3.4 Example of an ethological data sheet. In this example, one-minute focal sampling is being performed on an individual chimpanzee.

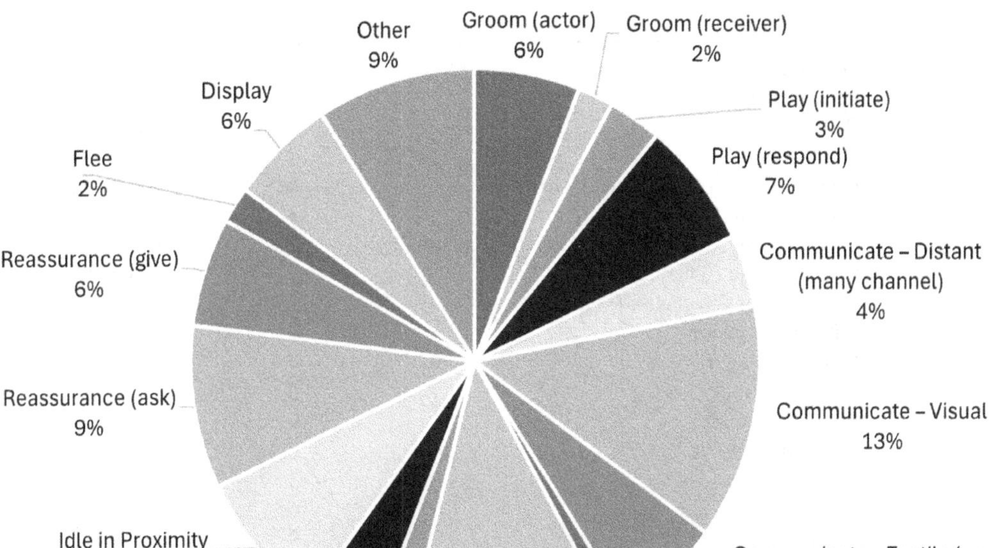

Figure 3.5 A sample pie chart of a breakdown of a social events within a social state.

In the case of our example, we can see that this individual spends the majority of his time communicating in some fashion (visually, vocally) as well as asking for reassurance and grooming others. He spends very little time initiating aggression or being groomed by others. We might gather from this that he is a lower-ranking, but very socially engaged chimpanzee.

We may, however, want to understand the significance of this data. For this, we need to employ a statistical test of significance that allows us to understand the actual occurrences of events versus what is expected. For our example, let's assume that, in addition to a percentage breakdown, we performed a chi-square test on our results. A **chi-square** is a statistical test that gives us a value that measures the distance between our actual results and expected results. Generally, if our value is above 0.05, we can assume that our results are significantly different from what is expected (either more or less than what is expected). In Table 3.3., we add a column with our chi-square value next to our percentage.

From this, we can determine that most events are in line with what we would expect from a normal distribution of social behaviours. However, we do see some

Table 3.3 A sample breakdown of behavioural events within a social state with chi-square value added

Event	Percentage	Chi
Groom (actor)	6	0.0081086956
Groom (receiver)	2	0.1037768744
Play (initiate)	3	0.0513934433
Play (respond)	7	0.0045533429
Communicate – distant (many channels)	4	0.0269954833
Communicate – visual	13	0.0001663506
Communicate – tactile (non-groom, non-aggression)	6	0.0081086956
Communicate – chemical	1	0.2419707245
Communicate – vocalise (non-play)	12	0.0002854648
Initiate aggressions	2	0.1037768744
Respond to aggression	4	0.0269954833
Idle in proximity	8	0.0025833732
Reassurance (ask)	9	0.0014772828
Reassurance (give)	6	0.0081086956
Flee	2	0.1037768744
Display	6	0.0081086956
Others	9	0.0014772828

areas where there are some significant derivations. For example, we see that the results from grooming, initiating aggressions, and fleeing are all significantly different (in these cases, lower) than what we would expect from a normal distribution. We may note that chemical communication is also significantly low, however, this is behaviour that would usually be outside of a normal distribution of behaviours for a chimpanzee and should be considered accordingly. Knowing such information is key to properly interpreting results.

If we combine all of our chi-square data on a visual plot, we can clearly see that the distribution of social behaviours is generally normal, and in the range of expected, for these chimpanzees (Figure 3.6.).

3.3 Physiological Measures of Thriving

Because behaviours are active, and most often, observable, they are an excellent tool for measuring an animal's current state with regard to thriving. However, behaviours cannot tell us everything. Many aspects of animal well-being are masked from behaviours. Therefore, in order to get the most complete picture of an animal's current state, behavioural assessments must be coupled with physiological assessments.

Consistent access to a robust veterinary programme is essential to the well-being of all animals living in captivity. Only with the direct performance and guidance of a fully trained veterinarian can physiological well-being be properly assessed and

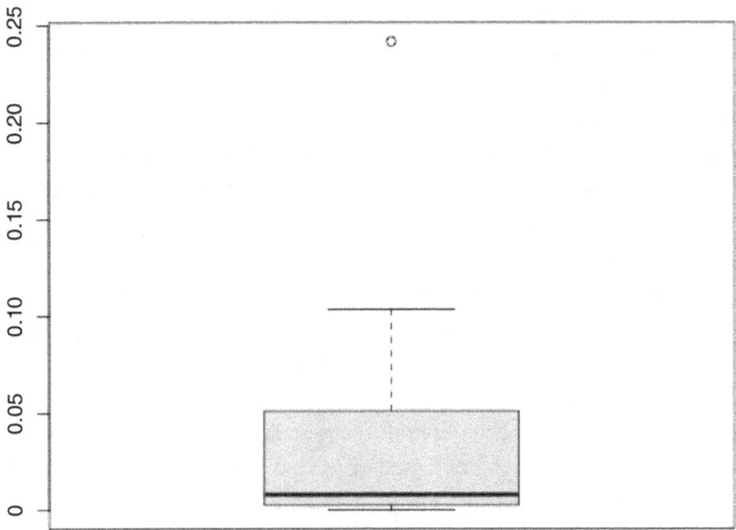

Figure 3.6 Sample chi-square as a box plot. Note the outlier at the top represents a behaviour that would not normally be seen in the range of a normal distribution.

treated. Key to this are veterinary rounds, whereby a veterinarian personally observes the animal in one's care and reports these observations to the care technician. This can lead to more extensive assessments and treatments being prescribed.

However, for a care technician curating a collection of animals, the responsibility is to ensure that all animals are receiving consistent and proper veterinary care; and to assist the veterinarian with physiological assessments.

Routine Veterinary Examinations

All captive animals must receive routine veterinary examinations. These exams can range from those that require anaesthesia, to simpler exams. These exams should include blood work, examinations of organ function, dental examination and teeth cleaning, and assessment of any physiological changes from a previous exam.

Written and permanent veterinary records are critical, and a care technician must ensure that all records are stored properly and are easily accessible. Records should be complete and written for those with no prior knowledge of the individual or their history.

Examinations should occur regularly. It is up to the institution's veterinary care to determine how often it is necessary.

Body Condition Scoring

Body condition scoring (or BCS) is a tool that enables a care technician (along with a veterinarian) to determine the relative health status of an animal by observing the overall state of an animal's body. BCS observations occur visually as well as (if safe to do so) through palpations of key parts of an animal's body to determine the amount of fat accumulation. BCS is a great method for examining an animal's nutritional status.

BCS works by numerical scale. The scale will depend on the species. However, generally, a very low score represents an emaciated individual, while a very high score represents an obese individual. Depending on the normal fluctuations within a species, scales can be as small as a 3-point scale, and as large as a 12-point scale.

It is important to remember that BCS is not just a tool to determine if an animal is eating well. BCS often denotes other problems an animal may be having. An animal whose BCS dips or raises unexpectedly may have a significant internal malady that needs to be diagnosed.

BCS can also denote behavioural issues such as anxiety or depression. An animal may not eat as much when depressed or anxious. BCS can let a care technician know how this lack of appetite is affecting the individual's overall health.

BCS functions by baselines. If an animal has a baseline BCS of a certain number, and that number changes abruptly (even if it changes to something more 'normal' within the species), the change may represent a problem that needs to be explored. For this reason, BCS needs to be performed regularly.

Many animals have a BCS that fluctuates seasonally. For these animals, it is important to perform multiple BCS assessments throughout the year. This way a care technician can derive seasonal BCS baselines.

Quality of Life

A **quality of life assessment** is an empirical tool to measure observable signs of well-being that collectively point to an individual's quality of life. Quality of life assessments can be species-specific or general. Like body condition scoring, quality of life assessments rely on a numerical point scale.

Quality of life assessments are frequently used to determine if an individual requires medical intervention; or, in the case of individuals with severe conditions, if euthanasia is advised. Whether or not the individual has a known medical condition, quality of life assessments should be performed on all animals within a captive population. They should also be performed consistently, and at regular intervals.

Below is an example of a general, multi-species quality of life assessment.

- *Example of a quality of life assessment*
  For each category, score the individual from 1 to 10.
  - o Mobility: Individual's ability to move about like a typical member of the species. (Note: this does not include environmental restrictions such as space. For example, flying space for birds and swimming space for aquatic animals. This scale is meant to measure only internal factors.)
    - ■ 10 – Movement is typical of this species.
    - ■ 1 – Little to no movement typical of this species.

- o Respiration: Individual's breathing rates as compared to those typical of the species. For instance, rapid breathing, shallow breathing, amount of wheezing, amount of coughing, and obvious breathing difficulties should all be considered.
  - ■ 10 – Individual breaths at a rate typical of the species without wheezing, coughing, or any sign of difficulty.
  - ■ 1 – Individual shows evidence of struggling to breathe. Welfare and activity are affected.
- o Excretion: Individual's ability, frequency, and health of urinating and defecating. For example, diarrhoea, constipation, struggling to urinate should all be considered.
  - ■ 10 – Individual excretes healthily.
  - ■ 1 – Individual's excretions show clear signs of a severe medical condition.
- o Diet consumption: Individual's appetite for food and water.
  - ■ 10 – Individual regularly consumes all prescribed diet and drinks an amount of water typical of the species.
  - ■ 1 – Individual is consuming little to no food. Water consumption is irregular.
- o Socialisation: Individual's engagement in social activity relative to both the species and the individual's baseline for social behaviour. Abrupt changes in socialisation should be considered.
  - ■ 10 – Individual is socialising typical of the species and typical of their individual baseline.
  - ■ 1 – Individual is engaging in little to no social activity.
- o Evidence of pain: Individual's visible display of any pain or discomfort. All evidence of pain should be considered (e.g., vocal evidence, limping, and self-examination of areas of pain).
  - ■ 10 – No evidence of pain.
  - ■ 1 – Evidence of severe pain.
- o Medical status: Individual's current medical prognosis.
  - ■ 10 – Individual has no known medical issues or conditions.
  - ■ 1 – Individual has a terminal condition and condition may be late-stage.

In this example quality of life assessment, a score of 56–70 would indicate overall positive quality of life, while a score below 35 would indicate a very poor quality of life. Anything between 36 and 55 should be a sign of concern.

Implications

Measuring an individual animal's ability to thrive is a multifaceted and scientific process that requires both behavioural and physiological observation and examination. The process can be time-consuming, challenging, and require a great deal of staff training. However, it is a crucial aspect of animal care that cannot be neglected, no matter the institution, facility, or environment.

References

Hill, S. P. and Broom, D. M., 2009. Measuring zoo animal welfare: theory and practice. *Zoo Biology: Published in Affiliation with the American Zoo and Aquarium Association*, 28(6), pp. 531–544.

Hinde, R. A., 1966. *Animal behavior: a synthesis of ethology and comparative psychology.* New York: McGraw-Hill.

Miller, L. J., Vicino, G. A., Sheftel, J. and Lauderdale, L. K., 2020. Behavioral diversity as a potential indicator of positive animal welfare. *Animals*, 10(7), p. 1211.

4   Barriers to Thriving

4.1   The Black Valentine

A white squall appears on the horizon. The crew mates on the deck of the Black Valentine exchange glances. They know that the end is near. Their wooden ship is no match for the typhoon they are navigating directly into. The captain knows that their lifeboats are useless. They will all go down with the ship.

The grey sky appears to be held up by thick waterspouts that pour water from the ocean up into the swirling clouds. This, combined with the lines of horizontal rain, make it seems as if the binary of up versus down has lost its meaning. Unable to shout above the roaring wind and rain, and unable to see clearly, the crew have lost their ability to communicate with each other. With its disoriented crew unable to communicate, see, or hear, the ship is unable to function. As the ship begins to take in water, The Black Valentine is no longer a ship. Rather, it is now just a piece of debris. As water completely consumes the ship, the crew are pulled into the dark sea. Darkness overtakes.

The captain opens his eyes. The sun is shining. The storm has passed. He looks around. He realises that he is on the shore of an island. He begins to remember the events of the storm. He stands up to look for his crew. Shards of wood from The Black Valentine cover the shore. He walks along to see if there are any survivors lying along the beach.

Walking along the lonely beach, the captain only sees pieces of his once proud vessel. As he walks further, he spots something. It is the image of a man sitting up beside the water. Frantically he runs to the figure, assuming it to be a member of his crew. Shouting as he runs, relief at the appearance of a companion fills his voice. As he draws closer, he sees that the figure is indeed a man; however, the man is motionless. He slows down. With each step, the horror of what is in front of him becomes clear. There, on the beach, is the contorted, lifeless body of his first mate. The deceiving position that the current placed him in, creates an eeriness that magnifies the dark reality. The captain cries out to the ocean. There is no reply. The captain is alone.

Hours turn to days. Days turn to weeks, weeks turn to months. Months turn to years. Throughout it all, the captain remains alone. During this time, the captain has manufactured some things on this lonely island that help him survive. With the washed-up wood and stained glass from the Black Valentine, the captain has constructed a cabin. The ship's large arsenal of kerosene provides fuel for the many lanterns that have also

Figure 4.1 Two newly introduced chimpanzees in a 'Special Needs' yard at Save the Chimps. (Photo credit: author).

come ashore. Other items that he finds also prove useful: pots and pans for cooking, knives for cutting branches and filleting fish, and a telescope to watch for any potential rescue. With an ocean teeming with fish, and an island forest full of edible fruits, the captain finds himself eating better than he ever has before. However, despite these benefits, the captain's new life is miserable.

Firstly, the captain is physically strained by his new environment. Life on The Black Valentine was a life of flat decks, maintaining balance as the ship rocked, and climbing rope masts. Over his many years aboard the ship, his body had become accustomed to this way of moving. The island is full of hills and large rocks. In fact, for the captain to access fresh water, he must ascend to the top of a hill where a clear pool exists. His knees pay the price and get so sore that just walking is painful.

The captain also finds the native wildlife to be problematic. The island is full of biting insects and venomous snakes. At night he can hear the guttural roar of a large animal he has not yet seen in daylight. The sound frightens him and fills him full of anxiety each night. To make matters worse, one night he is stung by a scorpion as he sleeps. For the next several days, he is immobile with a swollen leg and a metallic taste that fills his mouth. Each night after, he fears being stung again. Due to this, the captain gets very little sleep.

However, the worst torment the captain faces is the abject loneliness of his situation. His old world was, for better or worse, always social. He interacted with his crew to navigate the ship. He interacted with those on shore to do his business. He interacted with the regulars at his local taverns. There were also the people from his long-abandoned homeland. These were the people he missed most of all. As the captain sailed the world, he had left these loved ones behind. Now, fruitlessly, he wishes he had never left.

Before he had been shipwrecked, when the captain missed his home, he would stand on the bow of The Black Valentine and allow himself to feel surrounded by the vast ocean on all sides. He would feel comforted by the fact that the water he faced was the same body of water that reached all of those he had left behind. The water gave him a sense of being connected to them.

On one particular morning, the captain awakens to cooler weather. He has not yet experienced weather like this on the island. The crisp, cool breeze reminds him of his home. The chill in the air fills him with an even lonelier feeling. His mind is flooded with thoughts of his old world. He thinks of his dead crew. He thinks of those at home. He remembers holidays. He remembers Christmases. He remembers Valentine's Days. He wonders what the date is. Perhaps today's one of those holidays. He is immersed in the solitary reality of the loneliness of his new environment. He decides to do as he had always done on The Black Valentine.

The captain walks to the edge of the shoreline. He puts one foot in the water. The water is cold. He puts his other foot in the water. Slowly he walks out into the water until he is waist deep. He stares out at the sea in front of him. He waits to feel the connection to the rest of the world. In the distance, he hears a gust of wind approach. The wind blows over the surface of the sea and across the captain's face. The captain's long beard and hair billow in the gust as he closes his eyes. The piercing sound of the wind that has blown across an indescribable expanse, echoes over the miles of nothingness and is overwhelming in its emptiness. All of a sudden, the captain truly understands the vast distance between himself and his old world. The sound and feel of the great wind fill him with the emptiness of his lonely life. He is truly isolated from everything he has ever known. He will never be able to breach the chasm that separates him from where he once was and who he once was. All at once he feels isolation, regret, and hopelessness. His sore knees ache in the cold water. He turns back to face his forever island home. The dark forest on the shore is full of the mysterious roars, the stinging insects, and the snakes he fears. The captain curses the sea for sparing him. Why must he lead this existence? He looks to the sky and bellows a scream into the air. There is no one to hear.

Determining Obstacles to Thriving

The tragedy of the captain's situation shows us how complex obstacles to thriving can be. The ability to thrive is based on an individual's external factors (the environment), internal factors (physical well-being), species-specific needs, individually specific needs, and history of events. In the tragedy, we can note the following:

- *The captain is not thriving in his new environment.*
  - The captain is not able to thrive in this environment due to the wildlife, the terrain, and his access to resources.
  - The captain has two choices: he can either change his environment or acclimate to it.
    - Building his cabin was an attempt to change his environment; however, this fell short. Perhaps there are more changes the captain can make to his environment that he has not yet attempted. Perhaps this environment cannot be changed in a manner that will ever suit the captain's needs.
    - Years later, the captain has still not acclimated. Perhaps there are methods of acclimating to the environment that the captain has not attempted. Perhaps it will be impossible for the captain to ever acclimate to the environment.
- *The captain is not thriving as a human.*
  - As a human, the captain is a social species and requires companionship to thrive. In this environment, the captain is alone.
    - The captain attempts to mitigate this obstacle by walking to the water and imagining social companionship. However, this attempt falls short.
- *The captain is not thriving as an individual.*
  - The captain is not thriving on the island due to his own individual circumstances. These circumstances include his physical state and his psychological state.
    - The condition of the captain's knees is preventing him from thriving.
    - The captain's sleeping disorder is preventing him from thriving.
    - The captain's personal fears are preventing him from thriving.

- *The captain is not thriving due to history.*
  - The historic events that have created the captain's current circumstances have affected his ability to thrive as both a human and as an individual.
    - The natural event of the hurricane caused the shipwreck that caused the captain to become stranded in his new environment.
    - The event of the shipwreck killed the captain's crew and caused him to live without social companionship.
    - The memories of the captain's past history have exacerbated the captain's loneliness.
    - The events of having to traverse the rocky terrain and move up and down the hill have caused the captain to have physical challenges.
    - The recurring events of the unknown animal's roar in the dark forest have caused the captain to live with anxiety and avoid portions of the environment.
    - The event of the scorpion sting has caused the captain to lose sleep.

All individuals encounter certain obstacles to thriving. These obstacles can occur at the environmental level, the species level, the individual level, and at the historic level. More often than not, these obstacles are a combination of multiple levels. Within any population, there may be a host of obstacles to thriving. Some may be obstacles felt across the population. Others may be specific to an individual.

The Equation

As care technicians, we can mitigate these obstacles by examining the internal and external factors that may be at the centre of these challenges. First and foremost, we are able to provide an environment that is conducive to thriving and make changes to that environment when we find elements of it that we find may be detrimental. We can also become knowledgeable in the species we take care of by understanding what that species needs and what may be harmful to the general welfare of that species. Internally, we can understand the nutritional requirements and physical needs of the species. Externally, we can understand what the species requires environmentally and socially. Understanding the species is absolutely essential to knowing whether an individual is thriving or having challenges. It is also critically important to understand each animal as a unique individual with unique needs and unique challenges. It is essential that care technicians understand each individual's medical circumstances, each individual's social circumstances, each individual's behavioural circumstances, and anything else that might factor into an individual's ability to thrive. Finally, it is critical to know the unique major events that each individual, or the entire population, has faced. Each historic event may have affected an individual's ability to thrive at either the species level or the individual level. Knowing these histories may also help determine why a particular environment may not be conducive for a population or individual to thrive.

We can actually calculate the barriers and challenges to thriving with a simple formula:

$$b = \frac{e(s+i+h)}{m}.$$

In our equation:

- b = barriers to thriving,
- e = environment,
- s = species-specific variables,
- i = individually specific variables,
- h = historic variables,
- and m = mitigation efforts.

Through this, we can see that barriers to thriving basically amount to the combination of what a species needs, what an individual needs, and what that individual has historically encountered within their lifetime that has affected them physically or behaviourally. These variables are multiplied by environmental variables. These barriers can be significantly reduced (or eliminated entirely) by targeted mitigation efforts. It should be noted that changing any of these variables will significantly affect the outcome of this formula.

4.2   The Multiplier: Environmental Challenges

As indicated by our equation, the primary external factor inhibiting an animal's ability to thrive is the environment. Individual needs and barriers, species needs and barriers, and historic needs and barriers are multiplied by the environment. Thus, mitigating barriers and challenges to thriving must begin with a deep examination of the environment and how this environment may be exacerbating or creating these challenges.

As we will define in the next unit, the environment includes all external variables that create the living condition of an individual. Thus, the environment includes every dynamic outside of an individual's physical body. This includes terrain, air quality, water quality, noise, light, colours, shapes, odours, available food sources, the presence of other species and pests, the presence of observers, the presence of foreign objects, and any other outside circumstance that may be affecting an animal's general well-being. More often than not, these factors are able to be created, controlled, or taken away by the care technician.

When an animal, or population, is found to not be thriving in an environment, the care technician faces three choices: provide the animal or population time to acclimate to the environment, and any assistance that can be given to acclimate to the population; change the environment to a state where an animal or population can thrive; or move the animal or population to an environment where thriving can occur.

Acclimating to an Environment

A captive environment, by its very definition as an artificial environment, can be a source of chronic stress for an animal. This can be chronic physical stress (heart rate, raised hormone levels, etc.) and psychological stress (anxiety, boredom, etc.). Over time, however, the chronic stress can decrease based on acclimation.[1] Additionally, entering into a new environment requires so much adjustment for an animal that animals that are able to breed in a new captive environment show new phenotypic changes within one generation.[2] So that said, individuals will likely acclimate to an environment – even if that environment is poor. However, that acclimation may not always be positive. In fact, **negative acclimation**, or acclimation that creates chronic welfare issues, can be common in captive environments. Negative acclimation can be seen in examples such as abnormal aggression or stereotypic behaviours. So even though animals may have observable decreases in chronic stress, it may be getting replaced by negative acclimation behaviours.

When animals are introduced into a new environment, there is stress. Even in the best captive environment, the beginning of an acclimation process can be an extremely stressful event. Having the time to acclimate is always required. During this time, it is incumbent on the care technician to provide the tools for success to the animals, and assess the animals for their welfare to ensure that the acclimation is positive.

- *Assessing animal welfare during acclimation period.*

  All manner of assessment, both physical and behavioural, must be performed during an acclimation period. First and foremost, a dietary intake sheet is key to ensure that all animals acclimating to a new environment are consuming their required amount of nutrients. This should be coupled by regular body condition scoring to determine if any animal is gaining or losing weight unexpectedly. Concurrently, behavioural assessments, including performing care and welfare assessments and ethological observations, should be performed.

- *Providing enrichment that promotes engaging with the environment.*

  Enrichment that is designed to allow an animal to engage with the environment can be an extremely useful tool in aiding the acclimation process. Sometimes this can be as simple as distributing food in a manner consistent with natural feeding behaviours – such as dispersing food on the ground for foraging species. This allows an animal to explore their environment while being rewarded by the food that they find. Similarly, other food enrichment that makes use of standing structures within the environment can provide the same benefit. For example, browsing species can benefit from food being hung in novel ways from standing structures.

  Other types of enrichment that can aid in this acclimation process include adding novel scents to the environment, playing novel sounds within the environment, placing new textures within the environment, and even utilising new colours and shapes within an environment.

- *Utilising social introductions when appropriate.*

  When appropriate for the species, and, just as importantly, when appropriate for the individual, social introductions should be performed to give the individual what

---

[1] See Fischer et al., 2018.
[2] See Mason et al., 2013.

they require in terms of social companionship. However, it must be explicitly stated that some individuals may have preexisting physical or psychological challenges to being introduced to a conspecific. This is especially true with naturally aggressive or naturally territorial species. On the flip side, if the social makeup of an environment is different from what an animal would normally find in their natural ecology, it can have a significant negative effect and tremendous consequences for the acclimation process.

Solitary animals forced to live with conspecifics can be the source of undue anxiety and distress. Similarly, an unnatural sex ratio of a population can cause stress and/or aggression.

- *Making the decision to change the environment.*

After an assessment period has run its course, it may become clear that the individual or individuals are unable to acclimate to a new environment. At this point the care technician is faced with the decision to either move the animal to a new environment or change the environment significantly and accordingly. It is important to note that this decision can be made at any point during the assessment period. A care technician doesn't have to wait for the assessment period to run its course to make the decision that the individual, or individuals, will not be able to acclimate appropriately. Clear indicators that the animal(s) is not thriving, with the inability to thrive both psychologically and physically in the future, should prompt the swift decision to change the environment.

Changing an Environment

Changing an artificial environment so that an animal can thrive must be done according to both the species-specific needs of an animal and the unique needs of an individual. For the former, a care technician relies on the available scientific literature. For the latter, the care technician relies on evidence from previous individual assessments.

For example, if assessments are showing that an animal is not physically thriving through body condition scores, veterinary physicals, or dietary intake, the care technician must look at all the variables that may be causing such conditions. If the variable is shown to be environmental, this should dictate how certain aspects of the environment are changed. The environment may not be temperature appropriate. Food sources may not be enticing. Poor terrain may cause issues with movement. Any manner of changes may be attempted to treat such issues with physically thriving. New structures may need to be constructed. New environmental controls may need to be put in. New diets may need to be prescribed.

Similarly, negative behavioural assessments may lead to necessary changes in the environment. Social make up may need to be altered. More enticing environmental fixtures may need to be constructed. External sounds, odours, or even colours may need to be altered.

In both cases – behavioural and physical obstacles to thriving – the answers are usually very specific and unique. There is no universal approach to ensuring that changes can facilitate thriving. Oftentimes, this method is that of an educated trial and error where tweaks and changes lead to positive signs of thriving.

The 'Sordid Six': Key Environmental Obstacles to Thriving

While barriers and challenges to thriving are multifaceted and complex; and, as we have seen, frequently require a multi-pronged approach to assessing their causes and eventual solutions, there are six factors which will always create barriers and challenges to thriving. These 'sordid six', as we will call them, should be avoided at all costs. The sordid six are insidious in nature. They can befall even the best organisations. They appear as responses to limited resources, limited experience, limited education, and limited staffing. They get easily overlooked the longer they remain. However, they are anathema to good care and welfare. The good news is that they are all easy to fix once they are identified. Thus, frequent checks on the possible presence of any of these sordid six are paramount to proper care.

- *Environmental predictability*

  Environmental predictability becomes a part of the sordid six when what an individual animal perceives through their senses is significantly more predictable than the randomness that they would be naturally inclined to experience. Environmental predictability occurs on all levels of senses. What an individual sees, what an individual hears, what an individual tastes, what an individual smells, and what an individual feels should mirror the unpredictability found in a natural habitat. Common examples of this are dietary items and dietary presentation. Animals who are natural foragers should be permitted to forage at least part of their diet. Animals that are naturally inclined to eat a wide variety of foods should be given a variety of dietary items.

  Another example is landscape. Nomadic animals that are accustomed to experiencing different landscapes on a daily basis should be given a habitat that can be changed frequently.

  Knowing what levels of unpredictability an animal needs requires a care technician to be very knowledgeable about that species. Failure to avoid environmental predictability can lead to boredom and in turn, aberrant behaviours based on this boredom. A proper enrichment programme that targets eradicating environmental predictability is a panacea to eliminating this frequently stumbled upon member of the sordid six.

- *Restriction of natural behaviours*

  Restricting the natural behaviours of an animal is a massive barrier to the ability to thrive. This presents the care technician with a dilemma: by its very definition, an artificial environment is going to inhibit some natural behaviours. However, it becomes the responsibility of the care technician to provide the animals in their care with the ability to perform as many naturalistic behaviours as possible.

  Examples of this include a social species needing other conspecifics in their environment in order to have natural social behaviour; a browsing animal needing to be fed in such a manner as to allow the natural behaviour of browsing for food; a nocturnal species needing an environment where nocturnal activity can occur. As with others of the sordid six, the key to avoiding the significant restricting of natural behaviours is a keen knowledge of the species and the species behaviour.

There may be instances where an animal's natural behaviours will be so significantly impacted by an artificial environment, that serious consideration must be given as to the question of whether or not that animal should be in this environment or, rather, rehomed to an environment where the ability to practise more natural behaviours can occur.

- *Inappropriate environmental quality*

  Environmental quality includes all aspects of a habitat that might affect an animal's welfare. Environmental factors that can negatively affect welfare include sound pollution, light pollution, air pollution, uncomfortable housing, a lack of pest control, unsafe structures, dirty water, contaminated feeding surfaces, or any other aspect of an environment that might prevent an animal from thriving.

  Ensuring environmental quality means constantly testing and assessing the environment. Safety checks, water quality checks, decibel level checks, etc., are paramount to providing appropriate environmental quality.

- *Unsanitary environment*

  Perhaps the most obvious member of the sordid six is the presence of an unsanitary environment, the buildup of wastes, the buildup of bacteria and other microorganisms, the presence of animal pests and their wastes, and/or anything else that would create a lack of sanitation, greatly compromise an animal's ability to thrive in an artificial environment. Lack of sanitation compromises health, the ability to move around and general comfort. A lack of sanitation can also lead to rapid spread of disease and zoonotic disease spread between humans and the animals in their charge.

  Proper sanitation is easy to address with the presence of a proper substrate, a proper cleaning protocol, and proper tools and disinfectants. Arriving at a proper sanitation protocol is a joint effort that needs input from both care professionals and veterinary professionals to ensure that all manner of sanitation is effectively combating an unsanitary environment.

- *Improper diet*

  In addition to the thoughtful presentation of an animal's diet (see 'Environmental Predictability'), the diet an animal receives must be created to mirror what an animal would eat in the wild. All too frequently animals in captivity are fed processed food that bears little resemblance to what they would get in the wild. The results are nutritional deficiencies and an inability to physically thrive.

  Manufactured staple diets should be augmented with natural foods. Junk food 'treats' should be avoided. Herbivores should have a diet of produce. Omnivores should be given both meat and produce. Carnivores should be given meat.

- *Inappropriate social composition*

  Failure to provide the appropriate social companionship for an animal can have grave consequences. Solitary animals that are forced to live with conspecifics may live in a constant state of anxiety, or worse, become the victim of significant, or even fatal, aggression. Social animals deprived of companionship can become depressed, lethargic, or despondent.

Additionally, the sex ratio of an animal's social group should mirror that of the wild. For example, multi-male, multi-female social groups should be provided for animals that live with such social groups in the wild. Similarly, animals that live with only one adult member of one sex, such as gorillas, should live with such a ratio in an artificial environment.

4.3   The Multiplicand: Internal Challenges

As indicated by our equation, an environment multiplies the effects of the combination of species-specific challenges, individually specific challenges, and historic variables. When we look at the environment, we are looking at all of the external variables that an individual is faced with. However, in order to see how those external variables affect an individual, we must look at the internal variables that are ultimately multiplied by the effects of the environment.

Species-Specific Causes

The species-specific traits of an animal are the universals that provide the framework of who an animal is and what an animal needs. They exist as the evolutionary traits that define an animal physically, behaviourally, and ecologically. Internally, they allow an animal to thrive in a specific niche. Externally, they require the environment of that niche to realise these traits – traits that have been forged by the forces of selection over the course of the species' natural history.

Species-specific traits will determine every way in which an animal relates to its environment. They determine how an animal can move through an environment. They determine how an animal acquires resources in an environment. They even determine how an animal comprehends its environment. They determine what an animal perceives. They determine an animal's motivation. They determine every limit to an animal's world. As such, they determine the species-specific challenges to thriving that an individual will have in any given environment.

In an artificial environment, an animal has been given an environment they have not acclimated to over the course of their natural history. Therefore, artificial environments that replicate the species' natural environment tend to be best at accommodating species-specific traits and, thereby, avoiding many of the species-specific challenges an individual might have.

A mistake that is often made when attempting to replicate natural environments is to not look holistically at species-specific traits and only to see the replication from a very human standpoint. For example, an artificial environment may look to us like it is replicating the natural environment of another. However, on closer inspection, that habitat may not provide what the species needs. For example, there may not be adequate space. There may not be similar acoustics. There may not be the availability to move as they naturally would or use their limbs as they would in their natural habitat. Oftentimes, it takes observing that animal in this artificial environment to determine if this environment is adequately replicating their natural habitat.

Individually Specific Causes

Every individual of a species is unique with unique challenges, unique abilities, and unique references. These are an individual's specific traits. These individual differences may be determined by genetics, history, or rearing. Whatever the reason for them, they determine how an individual is able to thrive in a given environment. And as such, they dictate an individual's challenges to thriving.

Individual differences may be physical as well as psychological and behavioural. Different physical abilities determine different variables to the ability to thrive in an environment. Psychological differences can determine how an individual perceives an environment. Behavioural differences may determine how an animal responds to different aspects of an environment. In all cases, individually specific barriers to thriving have to be determined through close assessment and observation of an individual and can only be mitigated through the data collected in this assessment.

Something that might be difficult for a care technician is when an individual lives in a large population, and these challenges are unique to just that individual. In cases such as these, special care plans must be orchestrated to be able to treat this individual within the larger population, sometimes this is not possible. If this is the case and the challenge is significant, then that individual must be moved from that population and placed into an area where these challenges can be treated.

Historic Variables

Significant historical events can impact an individual or a larger population. Significant historical events can determine how an individual or population can thrive in a given environment. These events may be negative, or even traumatic, events that determine what an animal fears or what may give an animal anxiety. These events could also be conditioning events, dictating what an animal has grown accustomed to. These events may determine how an animal reacts to any given stimuli.

Obviously, as we have defined, an environment is the amalgamation of all external variables that an individual faces. Therefore, an environment determines what these given stimuli are. Care technicians are oftentimes hampered by the fact that it is not always possible to know what the significant historical events have been that have affected their population or individual animals. Once again, it will take a proper programme of assessment and observations to determine what the specific challenges are to an individual, which can then lead to an educated conjecture on what these historic causes may have been.

For example, an animal that was once in a poor quality captive environment that has now been moved into a proper artificial habitat may have episodes of displaying anxiety behaviour when food is present. There could be a number of historic causes to this food-based anxiety. We might surmise that this individual went through periods of food deprivation and therefore gets anxious when food is present. Or perhaps, we might conject, that this animal ate food that made them sick in the past and therefore gets nervous about eating. Any number of culprits could have created this challenge to thriving. Therefore, it will take a period of assessment and observation to make a proper conjecture to understand the causes of this challenge and the correct mitigation strategy.

4.4 The Divisor: Mitigation Strategies

If we look at our equation, we can mitigate barriers and challenges to thriving by targeting either the three multiplicands (species-specific traits, individually specific traits, or historic variables) or the multiplier (the environment). Obviously, targeting the environment to mitigate any challenges is going to have the greatest effect because the environment is the multiplier. However, this doesn't mean that all aspects of the equation can't be targeted effectively (the multiplicands).

Also, of note in our equation is that, since mitigating these challenges is a divisor, it is very difficult, if not impossible, to completely eliminate all barriers and challenges to thriving. Animals in an artificial environment are going to face challenges that they would not face in a natural environment. However, as care technicians, through proper mitigation efforts and proper targeting of what the barriers and challenges to thriving are, we can greatly and significantly reduce these challenges.

Mitigating the Multiplicands

Targeting the multiplicands can be very difficult. How can one change what a species has evolved to need? How can one change what has historically occurred with the species or individual? How can one change what an individual is physically, behaviourally, or psychologically? The answers are that we cannot change these variables. However, it may be possible to synthesise that which is needed due to evolution, individual uniquity, or history. Obviously, each case is going to be unique, and each case is going to be different. But oftentimes mitigating these variables takes the form of targeted enrichment and targeted operant conditioning techniques.

For example, animals that are nomadic by nature and may, in the wild, cover miles upon miles of habitat, have a species-specific need for miles of day range. How can such a need be replicated in an artificial environment? The short answer is that it cannot. However, it is possible that a robust enrichment programme that targets this need may be able to at least synthesise some of what the species encounters in a nomadic lifestyle.

Another example might be an animal that has historically been conditioned to eat at a certain time or after a certain cue (let's imagine that they were previously fed every time a light is turned on). Now they anticipate food anytime a light is turned on, even though this isn't their prescribed feeding time. This might be mitigated with an operant conditioning programme where the animal is conditioned to accept food in different lighting circumstances.

Mitigating the Multiplier

The environment multiplies the effects of all species-specific traits, individually specific traits and historic variables. As such, mitigating the environment is the most significant way to address barriers and challenges to thriving. Luckily, mitigating environmental causes to barriers to thriving is often the easiest way to mitigate challenges for a care technician. After all, it is likely that, of the variables that the care technician has the most control over, it is the environment (or all external variables that an animal is faced with). Adding elements to an environment, changing and amending environmental features, deleting environmental factors, or completely moving an animal from one environment to a more suitable environment, are all things that a care technician has in their wheelhouse. However, as we have mentioned on the previous pages, collecting evidence is key to figuring out exactly what in the environment can be changed, deleted, or added to; and when it is appropriate to completely replace an environment.

Implications

Determining how well an animal is doing in any given environment or if an animal is able to thrive at all in any given environment with respect to their species-specific traits, individually specific traits and historic variables is a chief responsibility for all care technicians. Key to this responsibility is gathering evidence in making informed decisions on how to mitigate any challenges and barriers to thriving.

As our equation lays out, an individual's ability to thrive in an environment is based on how their species-specific traits, individually specific traits and historic variables, get multiplied by the environmental effects with which they are faced with. In turn, this equation can give us an idea of what barriers to thriving an individual may have. Once barriers have been identified, and their root causes derived, care technicians can go about mitigating these challenges through targeting any of the variables that might be at the foundation of any challenges that the individual is experiencing.

It's important to note that even after barriers to thriving are identified and mitigated, it is the responsibility of care technicians to do continued assessments to ensure that new barriers and challenges to thriving are not arising, and if they are, to be able to treat any new challenge accordingly through the same evidence-based approach and through the same scientific methodology that was effective before.

Case Study: Animal Care in a War Zone. Working with the Kabul Zoo, Afghanistan

Brendan Whittington-Jones, Conservation Scientist

Brendan Whittington-Jones works for the Sharjah Government as a conservation scientist. He has assisted zoos in war zones, such as the Baghdad Zoo and the Kabul Zoo. He is the author of The Accidental Invasion of Baghdad Zoo.

Figure 4.2 Brandon Whittington-Jones. (Photo credit: Brandon Whittington-Jones collection).

On Getting Involved with the Kabul Zoo

I am a South African wildlife conservationist with degrees in wildlife management and zoology. In 2003 I worked as a tourist guide and game preserve manager in a small, private protected area in South Africa. Following the US Coalition's invasion of Iraq in 2003, my boss, Lawrence Anthony, and I decided we had the desire and a diverse enough skill set from managing chaotic, dynamic, wildlife-related situations with a limited budget that we may be able to assist animals in a zoo in Iraq. The inspiration for the idea was the memory of news articles covering a zoo rescue of sorts in Kabul following the US invasion of Afghanistan. What was meant to be my brief stay to help animals at Baghdad Zoo lasted for a year. At that time, North Carolina Zoo, under the leadership of Dr David Jones, had started assisting our Baghdad efforts with funding channelled from a network of sympathetic American Zoo Association sources. I left Baghdad Zoo with Dr Farah Murrani, an Iraqi veterinarian who had also volunteered at Baghdad Zoo. With an externship opportunity completed at Cheyenne Mountain Zoo in Colorado Springs, Dr Jones offered us new prospects at NC Zoo in 2005. While Dr Murrani was embedded in another externship, I travelled to Kabul to revive the rehabilitation work previously carried out by the NC Zoo in collaboration with the Zoological Society of London (ZSL), Mayhew Animal Home and WSPA. So, oddly enough, I left South Africa for Iraq in 2003, inspired by vague memories of Kabul Zoo, and two years later was standing in that zoo in Afghanistan.

On the Situation He First Encountered at the Kabul Zoo

Despite a year in the deteriorating context of Baghdad, my first thought entering Kabul Zoo was, 'What the f*** have I gotten myself into?.' The battered two-storey office building had eaten artillery strikes. A portion of the building had crumpled to rubble, and the dirty white plaster façade of the exterior was erratically pocked by shrapnel

Figure 4.3 Working at the Kabul Zoo. (Photo credit: Brandon Whittington-Jones collection).

hits from the civil war years. The British military had recently refurbished an office and a classroom among what was left of the building. Lush trees stood over and spilt out of areas with duck-filled ponds and shaded the tight network of paths. Visitor litter hid in the patchy grass and flowers and was liberally scattered along paths and among the bears and macaques. The heat and dust and smells were oppressive, in stark contrast to the photographs I had seen of the broken city blanketed in crisp winter snow. Acrid smoke billowed out from a partly collapsed, concrete-floored cage the size of a minivan. A keeper was burning the large bones leftover from the lion's food, to heat the water to boil the rice for staff, bears and birds.

The 4ha site was a Tetris-jumbled arrangement of walled paddocks, modified vulture and raptor aviaries from recent years, dirty, inadequately small holding cells for a myriad of animal species from birds to wolves, pigs and monkeys, a more exposed enclosure for lions (the famous Marjan had died and been replaced by then) and an elevated bear-pit of sorts holding both black and brown bears together. Visitors sat on their haunches on the empty moat's wall, laughing and tossing nuts at the bears that reared up and begged incessantly. A small mud island surrounded by a water-filled concrete moat was home to dozens of macaques playing out their social wildness among a few wooden platforms and log poles. Their holding rooms were dark rooms of faeces smeared walls partitioned with hair and faeces encrusted bars. Dozens of tortoises lay in a wet mud wallow while some ducks had no water. The former elephant enclosure was derelict, with the holding building slowly being repaired, a sign of the ambition to once again hold an elephant (the previous one had reportedly been gunned down by Taliban soldiers). Bullets and shrapnel holes decorated the brick and plaster.

Staff were polite, if initially somewhat hesitant about this new foreigner. Still, those that remembered the earlier work of Nick Lindsay and colleagues from ZSL were quickly hopeful there may be a positive outcome for them and the zoo. I was soon made to feel welcome with tea and quick smiles. It was increasingly evident that the neglected state of the zoo was more about a lack of organisation, inadequate resources and limited and outdated knowledge than reluctant attitudes or poor work ethic. Some staff would prove to be proud, incredibly hospitable and kind, slogging away daily with little reward other than an inherent sense of purpose. There was almost no awareness of what sort of state was acceptable in a modern zoo. Nobody

Figure 4.4 Brandon Whittington-Jones assisting the staff. (Photo credit: Brandon Whittington-Jones collection).

had a reference point; the surrounding city was in a state of filthy derelict chaos, and staff barely earned enough to keep their families alive. In context, the zoo wasn't unusual. In fact, it was popular, busy in the afternoons and crowded on weekends. The entry fee was negligible, much like the visitor discipline that the keepers had long since stopped intervening with. While women would stand motionless in thick blue burkas, there were always a few men who would shout and laugh at the animals they were teasing with headscarves or food packaging. There was room for improvement.

On the State of Animal Care at the Kabul Zoo

Animals were surviving. Unless they fell ill, then they usually died. There was adequate and varied food provided through funding by the Kabul Municipality and the assisting NGOs, and NC Zoo. The physical condition of most animals was generally surprisingly good. The mental states of many were unsurprisingly poor, with almost all canids, bears, some birds and caged monkeys showing apparent stereotypical behaviour. Despite this, keeper staff arrived at work each day, immediately set to rudimentary cleaning of animal facilities with what little equipment they had and fed and watered where appropriate. Most staff had a genuine interest and pride in the animals under their care but little insight into what modern standards were acceptable.

The veterinarian was almost always available, humble-natured and willing to get involved. Unfortunately, he lacked the skills and knowledge required. His clinic was two small, dark, musty, derelict rooms with large decaying taxidermied animals and zoo tools partly contained by broken cupboards. He only had access to a few generic medications he knew how to apply. However, he feared being injured in many situations and left medicine applications to the keeper staff.

External parasite loads on the animals, particularly ground birds crowded in small, heavily meshed cages, were high. Modern anti-parasiticides and vaccinations available in modern zoo facilities were unfamiliar options. Despite the willingness to do so, the keeper staff had little training on making observations that could indicate health concerns. More food and vitamin powder were usually the treatment for any health problems. As with many zoos struggling to modernise, enclosures and aviaries were typically small, and sparsely decorated and, although they were cleaned daily, were rarely clean enough. Food was for consumption, and the concept of stimulating foraging or other natural behaviours was unheard of. Enrichment did not exist.

Perhaps the starkest example of the knowledge deficit impacting husbandry were the snakes donated by visitors over time. The prized specimen, a species I unsurprisingly didn't recognise, was left outside in a birdcage with a boiled egg and a saucer of milk. The rest of the collection was housed in a single sizeable indoor terrarium with only a substrate of sand in a building with no electricity. They were not provided food or water but the occasional boiled egg and saucer of milk for six months until I arrived. The only light was through slits in the back of house doors of outdoor aviaries in the same corridor. Naturally, I was baffled to find one snake still alive as we sifted through the terrarium sand.

As for the state of the zoo, it was not only a hesitation to act towards improvement that was lacking; it was outdated guidance rooted in an understandably rigid survival mentality combined with cultural value judgements. In some cases, this was reflected in animals that may be euthanised in modern zoos but were kept alive until they died naturally. Killing a zoo animal was considered unacceptable by the zoo management's religious beliefs (with an added fear of official municipality judgement). It reflected the many welfare challenges that would need to be navigated through long informal discussions and months of building trust.

On the Dire Situations He Witnessed

In terms of the desperation of animals in inappropriate enclosures, the wolves and foxes were the worst off. Several grey wolves lived in about 25m² without shelter from public view, while numerous foxes lived together in cages even smaller and with little access to sunlight. A Pallas's cat lived in a cage likely not more than 4m². Those animals lived in states of perpetual stereotypic pacing. Again, it must be stressed that this was the product of the facilities at hand, designed according to old European standards with modifications based on a Soviet design blueprint with all text in Cyrillic script.

On an occasion, a pet rhesus macaque was donated to the zoo by an Afghan visitor. Although the gesture appeared to be one of sincere generosity, a familiar scenario, the question remained whether this young female would have to live in isolation or could be integrated into the existing social melee of the crowded island. It was agreed that after some time in a secure holding area within the larger macaque indoor housing, where social exposure could be allowed but keep the new female physically safe, the female would be released onto the island. After weeks of managed interactions, the female was released. She was chased, bitten and then actively drowned in the moat by an aggressor until keepers could intervene to rescue her. The macaque survived another day until zoo management reluctantly agreed that I could be issued a drug to euthanise her, provided I took responsibility for whatever otherworldly ills may befall me.

Perhaps a final example, one that reflects the remarkable context of the zoo and its staff, was the escape of a large brown bear from its enclosure into the zoo early one evening. The story was casually recounted to me the following morning with amusement of how the staff with flaming sticks corralled and then herded the bear back into its enclosure before securing the lock.

On How He Went about Assisting the Kabul Zoo

As the only foreigner on-site, I quickly relied on the English competency and willingness of the NC Zoo-sponsored Education Officer, Faizal Saidal, and the zoo Deputy Manager Shah Nuri, as day-to-day working partners and cultural guides. While understandably initially hesitant about my presence, the staff quickly understood my arrival was a continuation of the welcomed assistance by Nick Lindsay and his ZSL colleagues years earlier. Significant changes in infrastructure were not possible, both from a funding perspective and the zoo management's apparent reluctance to deviate from the old Soviet blueprint that was considered the official government layout. I advocate early, easy wins to build morale, mutual trust and momentum.

Many early wins were inexpensive, but demonstrated improvements were more about sustained intent and planning rather than only finances. Large barrel bins were painted brightly and distributed through the zoo for litter. Staff cleared all litter from the site. Broken taps and piping were repaired. Where there was no water supply but needed to be, these installations were quickly enabled. This allowed for holding facilities to be scrubbed down and disinfected. It enabled a more effective cleaning routine to be reinforced. Electricity and water pump systems that had fallen into disrepair were fixed. A large storage room in the main building was scrubbed down by staff, as food previously piled haphazardly over the floor was organised into containers and shelving to reduce wastage, better manage quantities and limit vermin damage. With a regular supply from a local market, a gas cooker system was installed to stop the outdoor burning of bones for rice cooking and enable more comfortable commissary management during the winter cold and summer heat.

For a few hundred dollars for a contractor and effort by the keeper and management staff, the veterinary facility was repaired, painted and organised into a working space. A veterinary NGO and Italian military veterinarians in Kabul assisted with donations of routine medications to build a small, pragmatic stockpile. The emphasis on preventative veterinary care partly mitigated the knowledge and skill deficit for acute and chronic reactive treatments. Where appropriate, vaccination routines were established, always an exciting prospect when a pole syringe is the only delivery option and bears and lions aren't familiar with routine confinement. Anti-parasitic medications were applied where possible. Since dietary routines were well established, familiar to staff and tailored to locally available produce and meat, that remained unchanged.

Although these were primarily cosmetic or rudimentary improvements across the zoo, everyone involved needed to be comfortable that the changes were noticeable and mostly sustainable without foreign support. With the trust established that modified daily routines would be easier to fulfil and be more effective than previous approaches, keeper staff were receptive to modifications and enrichment tests inside the enclosures.

The substrate was an easy target for experimentation. With sand and soil delivered to at least partially cover concrete enclosure floors, keepers could immediately see behavioural responses like birds scratching the ground and dust bathing. Porcupines excavated deeper soils. Keepers soon took small steps of initiative, tried to plant grass and started to take pride in testing out small changes in settings for animals. Basic dialogue about animal welfare and behaviour started. With some convincing, the macaque island was adorned with more ropes and logs. Food was scattered for bears to focus more on foraging and less on ripping up their concreted island. With these small changes, keeper staff started to be slightly more reactive towards visitors harassing animals. With that in mind, we facilitated some barricades to reduce direct interaction between visitors and the bears and macaques on their respective islands.

While routines, staff and animal welfare were improving slowly, there was a renewed push to focus the education programme on improving signage and proactively engaging schools and orphanages. It was an exercise in patience and expectation management for both the Afghan staff and me. It was also a reality check on the cultural complexities and the post-Taliban social dynamics and local politics, particularly as it related to gender and inter-agency permissions.

Over numerous visits, the common thread was to ensure that 'the basics' were incrementally done better. Recognition of context and limitations was crucial. Any changes needed to make logical sense to the Kabul Zoo staff and ideally be entrenched in routines that would remain unchanged even as management staff were redeployed to other municipality departments. Prudent use of funding was a daily challenge rooted in the question of 'is this better spent to help more animals less, or fewer animals more?'

On the Ultimate Outcome of His Work

I can't be the impartial judge of that. As much as I desired to initiate substantial infrastructure and welfare changes for animals, that was unrealistic. There were tangible changes in improved effectiveness of daily animal care routines, logistics and facilitating a slightly more extensive network of supporting foreign entities in Kabul. Upon return visits to the zoo, it was noticeable that keeper staff and management were sustaining the improvements. The objective was far greater than my contribution, as seen in the momentum demonstrated by the continued financial and educational support to Kabul Zoo by organisations like NC Zoo, ZSL, Mayhew Animal Home, and later the South Asian Zoo Association for Regional Cooperation (SAZARC) and Wildlife Conservation Society (WCS).

On If This Work Could Inspire Other Animal Care Initiatives in Unstable Regions

These support initiatives do not occur in a vacuum by individuals. Organisations like the NC Zoo, Zoo Outreach Organisation (ZOO) and Wild Welfare all have the expertise and, at times, access to resources to enable sustained positive change. I don't advocate for or support poor quality zoos and would prefer there were fewer to no zoos in unstable regions. However, if there are substandard facilities in unstable areas that are willing to engage transparently to improve conditions for staff and animals, I have no doubt there will be NGOs ready to assist.

References

Fischer, C. P., Wright-Lichter, J. and Romero, L. M., 2018. Chronic stress and the introduction to captivity: how wild house sparrows (*Passer domesticus*) adjust to laboratory conditions. *General and Comparative Endocrinology, 259*, pp. 85–92.

Mason, G., Burn, C. C., Dallaire, J. A., Kroshko, J., Kinkaid, H. M. and Jeschke, J. M., 2013. Plastic animals in cages: behavioural flexibility and responses to captivity. *Animal Behaviour, 85*(5), pp. 1113–1126.

Part II

External Factors

Curatio Fundamentorum **4:** An individual's environment is a combination of the following: a habitat; the degree of physical movement possible; the degree of mental stimulation present; the social dynamics with conspecifics; and the degree of intrinsic motivation afforded. The variables of these external factors determine an individual's ability to thrive in an environment.

Interlude: Chimpanzee Introductions

The sound that a slider-door makes in a chimpanzee holding area can fill me with all sorts of emotions at once. It can fill me with anxiety and worry. It can also fill me with excitement and awe at what happens once those doors are opened. Such was the case when one such door was opened in front of a chimpanzee named Clay.

Clay

Clay was born in 1987 at the Coulston Foundation in Alamogordo, New Mexico. At just 10 hours of age, Clay was pulled from his mother and placed into a nursery facility. For the first 12 years of his life, he was used in an array of studies; many of them extremely invasive. During this time, he had extremely limited social experience with other chimpanzees – spending most of his life completely alone.

Early in his life, Clay's care technicians had attempted to introduce him to other chimpanzees, but these introductions had gone wrong. At that time there was very little ways of set methods of performing chimpanzee social introductions in captivity. Add to this that a chimpanzee, such as Clay, with very limited social experience, would have required a very particular approach; both for his safety, and the safety of any other chimpanzee he met. Struggling with this, as well as a general lack of experience, his care technicians had attempted an introduction which resulted in injuries to both Clay and the older female he was introduced to. Eventually, the older female died after the introduction. Unfortunately, Clay was viewed as the one responsible for her death, and this stigma stayed with him. Since that time, Clay had lived a solitary existence, with only the ability to see other chimpanzees through mesh and hear the sounds of chimpanzees interacting with each other. Sadly, these were sounds that Clay could not participate in.

Because of this, Clay's introductions needed to be treated with extreme care. This wasn't necessarily because I was afraid that Clay would become too aggressive with other chimpanzees. Actually, I was afraid that Clay would not understand the consequences of his actions with other chimpanzees. I feared this because, after living alone for 20 years, it was likely that Clay would have no concept of how to interact with other chimpanzees. Therefore, I was afraid that Clay might not be able to understand what to do if conflict should arise or how to resolve any conflicts that did occur. Before he had a full introduction, we began with what we called 'protective contact' introductions. These were sessions where Clay was able to 'meet' other chimpanzees by being mesh-to-mesh with them. Through the mesh, the chimpanzees could touch each other, groom each other, and even tickle and play with each other. However, they were largely protected from any interactions that could become problematic.

After repeated protected contact sessions, we felt confident in Clay's ability to *fully* interact with another chimpanzee. At that point, we opened up a slider-door and fully introduced him to another chimpanzee. The first introduction (with a fairly dominant male) didn't go so well. A conflict arose that continued without abating. Our initial concerns were validated, and Clay didn't seem to know who to end or resolve the conflict. Thus, when they threw the slider-doors, we quickly shut them and separated the two chimpanzees.

Even though this first introduction didn't go as well as we hoped, there were signs of hope. We saw an acknowledgement that Clay did understand a bit of how to inter-act with another chimpanzee. We even saw signs that Clay was beginning to submit to a more dominant chimpanzee – however, he just didn't quite know how. This gave us the encouragement to continue.

We began introducing Clay with other chimpanzees. This time, we chose chim-panzees who expressed more diverse social roles. We introduced him to a dominant female, a submissive female, a submissive male, and a dominant male. Each time, Clay seemed to get a little bit better at understanding his own social role.

Jeff's Group

While Clay's introduction plans were being carried out, the sanctuary acquired seven chimpanzees from a facility that had run out of money. The facility was in fairly poor condition. The chimpanzees there had lived with limited social interactions, limited enrichment, and limited space. Of the seven, two were housed in one group, while five were housed in another. The group of two was fairly quickly able to be integrated into a larger group. However, the group of five was a bit more complicated.

This group had been together for over 20 years. It consisted of one adult male named Jeff, and four females. What made this even more complicated was that Jeff was both geriatric and extremely dominant. The four females were extremely closely bonded with him. There was Shake, a large female, who was his closest companion. There was Ernesta, an even larger female who was extremely dominant. There was Jeff and Ernesta's offspring, Magic – a very active, very intelligent, and fairly mis-chievous young female. Finally, there was Vanilla, Shake's sister, who was a bit of a loner. This group, with its bonds over time, had deeply ingrained social roles, and it was going to be challenging to integrate them into a larger group.

This was borne out by our experiences. Our curator, Ashley, spent tireless efforts creating introduction plan after introduction plan. Each time, Jeff would sit in the background while the females would not let any other chimpanzee near him. With other chimpanzees being unable to make any contact, Jeff wasn't really able to meet anyone. This would take some ingenuity.

A Plan

Our first thought was to temporarily separate some of the females from Jeff and allow them to create bonds with some other chimpanzees. We did this first with Vanilla and

Shake. They were introduced to one of our very established groups. Without Jeff, these introductions went very well. Vanilla and Shake got along very well with this group. Soon we were able to open up the slider-doors for them to go outside and step onto their large island habitat for the first time. It was quite a moment for all of us when Vanilla, who was first afraid to step out on the island, got encouragement from her newly introduced alpha male, Dwight. She found the courage, stepped outside, and looked up at the sky for the first time in her life.

Though Vanilla and Shake were doing well, we still had the issue of how we were going to incorporate the very dominant Jeff into the group. Even more concerning were Magic, who herself was very dominant, and Magic, who seemed to thrive on starting conflict.

One morning the curator, Ashley, barged into my office with a novel plan. Because Clay was just learning his social roles, and the chimpanzees in Jeff's group were so deeply ingrained in their roles, what if we used the disparity to everyone's advantage? What if we piggybacked Clay with Jeff's group and fully integrated them all into Dwight's larger group? Jeff's group was going to have to completely adjust their social roles. In turn, Dwight's group was going to have to adjust to allow the social space for Jeff's group to come in. This period of social evolution for both groups could allow Clay the space to find his own role within the group. I liked the idea. Ashley began to create a plan.

Chimpanzee Introductions

Like all chimpanzee sanctuaries, we were saddled with the care and well-being of a very large number of chimpanzees with a finite amount of space. Not only does proper chimpanzee welfare dictate social introductions, *logistics* dictate social introductions.

However, chimpanzee introductions can be very unpredictable, and even dangerous to the chimpanzees. As such, they must be approached in a very scientific manner and according to a plan. Chimpanzee introductions are never easy, but the easiest choice is not the best choice for chimpanzees who live in managed care. Thus, to ensure that the sanctuary always follows through on what is necessary rather than what is easy, we established a philosophy of care to ensure that we meet those standards.

A key component of that philosophy of care is that the sanctuary will always facilitate resocialisation and integration – with the recognition that chimpanzees are individuals with individual challenges and individual needs. Indeed, some have long periods of social isolation, limited space, and possibly even trauma. These chimpanzees might not adapt well to living in large social groups. This, we believed, was the case with Clay. It took a lot of planning for us to get to this point.

Every chimpanzee, regardless of their individual challenges and their individual needs, deserves as rich a social environment in so far as their individual abilities dictate. This, we must provide, along with freedom of choice and a cognitively stimulating environment. A rich social environment is foundational to behavioural thriving in a chimpanzee, and it is our job as care technicians to find each chimpanzee's social threshold and work with that, to bring them as close to thriving as possible without causing undue stress or injury.

The reason that chimpanzee introductions can be so dangerous and unpredictable rests in the dynamics of their natural ecology and natural behaviours. Chimpanzees have very definitive hierarchies, and they function on both geographic and social territoriality.

Geographically, chimpanzees have defended territories within their home range. This is what's known as their core range. However, in a managed setting (in an artificial environment), all areas of a habitat are core range. This means that every coroner of an artificial habitat is defendable, and by nature, they will defend this core range against any newcomer.

Additionally, chimpanzees have social territory. Chimpanzee hierarchies have a fixed membership that's specific to the group's dynamics. In an artificial environment, hierarchies are even more fixed due to the lack of ability for natural emigration and immigration. So if a foreign chimpanzee enters this social territory, there is likely going to be conflict, and this conflict is usually physical.

Determining Individual Social Threshold

Complicating all of this is how unique each individual chimpanzee is. This leads to extreme unpredictability for chimpanzee introductions. Each individual has their own social threshold. So each chimpanzee that participates in an introduction is going to react in their own unique way – and influence the introduction uniquely.

So, when creating a chimpanzee introduction plan, the first thing to do is to determine an individual's social capacity. This can be accomplished by looking at an individual's past social history and current behavioural variables. These variables might be dominance, submissiveness, social anxiety, a propensity for aggression, or any other relevant behavioural variable. The other thing that's going to determine a chimpanzee's social capacity is their health status. Are they healthy enough for the stress, both physical and psychological, that an introduction is likely going to induce?

Determining a Group's Social Capacity

After assessing individuals, one needs to look at a group's social capacity. A major determining factor is going to be the population size. The population capacity is going to be specific to that group based on hierarchy and the group's stability. However, it is important to be careful with this, though because a *stable* group is not always the *best* group. In fact, stability often means that the group's social territory is maxed out, and adding another individual can irreparably compromise the existing hierarchies and social stability. A care technician must look at the dominance hierarchies that are present within the group to see if there is a 'hole' for the new individual to occupy. This demands looking at the existing sex ratio of the population, the amount of dominant individuals, and the amount of submissive individuals.

Utilising Social Roles

A major factor in why we've been able to be very successful at chimpanzee introductions, is our method of utilising *social roles* in determining our introduction plans. Social roles for chimpanzees go beyond just looking at dominance and sex. Like humans, chimpanzees carry within their population very defined social roles. Those social roles fulfil a vital function within the group. One can almost think of these as each individual's niche within the functionality of that group. Like an ecological niche, it is difficult for chimpanzees with identical social roles to occupy the same space.

As such, determining social rules is absolutely critical to a successful introduction with chimpanzees. Alpha, mediator, outsider, loose cannon, elder statesman, and so forth, are all possible examples of social roles that can exist in both the wild and within managed populations. So when creating an introduction plan, I like to identify these social roles for each individual that is participating within the group. However, what is a very important caveat to all of this, is that as soon as the introduction occurs, those social roles are likely to change. Even so, understanding the social roles before going into the introduction, can give a bit more predictability to how that introduction is likely to go. It's all a bit like looking at jigsaw puzzles and seeing which puzzles have missing pieces. Finding chimpanzees who can fit into these spaces gives the best chance of success for an impending introduction.

Methods of Chimpanzee Introductions

There are plenty of workable methods to successfully perform a chimpanzee introduction. However, the key is in choosing which of these methods best fit the given circumstance. Each introduction is unique, and though there may be a preferred method, sometimes these unique situations dictate utilising something different.

The three primary methods are group building, introduce-and-separate, and coalition building. However, more often than not, it is a bit of a hybrid between two or all of these methods.

The first is what we call group building. This is when a core group is created with new individuals. Then, chimpanzees from the destination group are added (usually no more than four at a time). The order is determined by social rank, social role, sex, or any other relevant factor. This new group settles for at least 24 hours before adding other individuals. During this time, behavioural observations and assessments are performed to ensure that the group is stable enough to add more individuals.

The next method is known as introduce-and-separate. This is also sometimes called 'play dates'. In this method, individuals are introduced, allowed to spend a period of time together, then separated. This method might be the most workable when any of the chimpanzees involved in the introduction have significantly limited social experience.

Finally, there is coalition building. This occurs when small subgroups are created (through either a group-building process or an introduce-and-separate process) to form multiple coalitions. Once formed, all of the new coalitions are brought together. The success of this method hinges on creating coalitions that are equally matched and equally balanced so that one coalition doesn't overpower the other.

It's very rare that I've chosen just one method. Though my preferred method is group building, I usually hybridise this method with the other two in some capacity. This process has led to over 250 chimpanzees at the sanctuary being successfully integrated with each other.

Writing a Plan

Every chimpanzee introduction must proceed from a written plan. Though introductions should be flexible in process, this flexibility should be written into the plan. Once written, the plans should be reviewed by a committee of multiple care technicians and even those from different departments, such as veterinarians.

Written plans are critical for the simple fact that they enable a review process and ensure a planned strategy, rather than a reactionary approach. The moments of a chimpanzee introduction can be scary moments for the staff performing them. The best tool at their disposal is a written plan that includes an order of events and contingencies should things not go as planned.

At times introduction plans must be amended. However, when plans are amended, a written amendment to that plan should be created. Like the plan itself, the amendment can then be reviewed by the same committee that reviewed the original plan.

The anatomy of a chimpanzee introduction plan includes the 'how' (how is the plan going to proceed), the 'who' (which chimpanzees are included, and at which time), the 'where' (where will the introduction take place), and the 'why' (justification for why this introduction needs to occur).

The 'how' portion of the introduction plan should include which method is being utilised. Will it be group building? Will it be introduce-and-separate? Will it be coalition building? Will it be a hybrid? It should also include what would need to occur for the introduction to be shut down. Are minor injuries acceptable? Is there a chimpanzee with a health concern that would warrant shutting the introduction down should they show signs of this concern? The introduction plan should also determine which staff can be present during the introduction. The introduction plan should include anything else that is relevant to how the introduction will proceed and what would cause the plan to change or shift. Therefore, all possibilities should be included (as much as possible).

In the written plan, the 'who' portion contains all the individual chimpanzees that will be factored in each phase of the introduction. Some plans may be more flexible than others in this regard (however, even this flexibility should be written into the plan).

The 'where' portion dictates where the introduction will take place, what space is being utilised, and what swing space, if any, can be opened up for the chimpanzees if needed.

The 'why' portion of the plan is, perhaps, the most critical. Why is this introduction worth the risk to these chimpanzees? This should be clearly noted and clearly spelt out within the plan.

Once the Door Is Opened

As a care technician for chimpanzees, you quickly realise that any control you have over the actions of the chimpanzees in your care is largely an illusion. It is ultimately up to the chimpanzees themselves to act according to what you may have planned. Each time the slide door in front of Clay was opened, it was with this knowledge that we waited to see what occurred. However, what we could control was our plan and our execution of that plan.

Clay's plan worked. When he was introduced to members of his new group, he quickly assumed a role we had hoped he would. At the time of this writing, Clay lives in a group with 22 other chimpanzees. He has 24-hour access to a large 3-acre island habitat. He has the freedom of choice to spend time with any member of his group, wherever in the habitat he wishes. He is forming bonds with companions that he can communicate with, groom with, play with, and yes, even fight with. These companions can bring him solace, stimulation, and enrichment. None of this could have been done without utilising a scientific approach to chimpanzee introductions.

5   The Environment

Figure 5.1 Mixed species exhibit at the Kansas City Zoo Aquarium. (Photo credit: author).

5.1   The Magic Gin

A priest enters the public house. He is here to see The Mariner sitting alone at a table in the corner. The priest walks to the table, pauses for a moment, pulls up a chair and sits down. The Mariner looks up from his tankard of ale.

'I was told that you sent for me', says the priest. 'I was told that you wanted me to hear your confession. I am happy to hear your confession, but this pub …'
'Please Father', interrupts The Mariner. 'This is the only place I feel like myself.'

The vicar sighs. 'Very well', he says. 'Let me hear your confession'.

The Mariner takes a sip of his ale. He wipes his mouth, straightens his back, and begins. 'All I have ever wanted was to be a good person. My entire life, I have fallen short of that goal. Selfishness, bitterness, fear, and ego, have kept me from being that. I have accepted that. However, what I also realise is that – not only am I not a good person, I haven't even done right by the people that I love. I haven't even done right by the people who have shepherded me through life. This is something I cannot abide by. This is something I cannot live with.'

The priest smiles. 'This is something that is normal to feel. The fact that you recognise this and feel remorse proves you a righteous …'

'But there's more, Father', interrupts The Mariner. He looks down into his ale. He looks as if he's going to say something, but he stops himself. Finally, he bellows, 'All I wanted to be for her was a positive force in her life, an absolute good. I wanted to be for her what she had always been for me – a sanctuary from the world. I wanted to be the person she turned to when all else had failed. I wanted to be the person that brought her peace when she faced the worst. I was not that for her.'

The priest puts his hand on The Mariners' arm. 'Tell your confession.'

The Mariner straightens his back and begins to speak. 'I was once a sailor. I travelled along the southern trading routes. I travelled to the parts of the sea where there is no winter. I would sail from island to island, trading goods and making a living. I came to know a village on one of the islands very well. They knew me. They expected my arrival. I watched their children grow. I watched their old perish and die. It was a wonderful relationship, until I fell for her.'

The priest furrows his brow. He beckons the barmaid. She brings him an ale.

The Mariner continues. 'I had known her since she was a young woman. In time, I began to fall in love with her. She, in turn, fell in love with me. My heart would break when I had to leave the island and return home. After years I had finally had enough. When I returned to the island, I told her that she had to come back with me. She resisted, saying that the life on the island was the only one she knew. She spent her days capturing fish in the sea. She spent her nights around a fire. She could track the changes of season by the amount of rain. She was an important part of her village. She learned from them and, in turn, she taught what she knew to the young.'

'I told her that if she didn't come with me, I would never return again. I would never bring goods to her village. She would never see me again. Finally she relented, climbed aboard my vessel, and sailed with me to the north.'

'Once there, it was clear that she was not happy. She could not adapt to the cold winters. She spent the time shivering in a blanket beside the fire. She could also not get used to the food. Instead of the fresh tropical fruits she had always lived with, all I could offer her was the bland fruits of our country. Instead of the fresh fish she ate, I could only offer her fried cod. She missed her village. She had no common experiences with the people in my town. They also looked at her as if she were a perpetual stranger.'

'However, what appeared to be the biggest loss for her, was the loss of her lifestyle. At her village, she had developed keen skills to catch fish in her part of the sea. She had developed an expertise in where to find the best fruits on the island. She knew every part of her island. She knew where the dangers were. She knew where it was safe. She knew how to traverse the rocky cliffs, and how to cross the swampy valleys. In her new home, she could not utilise any of these skills. She had no areas to fish. She has not fruited forests. She had no rocky cliffs. She had no swampy valleys. Instead, she was thrown into a land where there were new challenges that she had never experienced. She found that she lacked any of the skills necessary to survive this foreign land. As a result, she wavered back and forth between extreme boredom and extreme anxiety.'

'When she told me that she could not survive here anymore, I became enraged. I told her that it was her duty to make the best of it. When I caught her trying to send a message to the people in her village, I became filled with jealousy. I accused her of loving the people in her village more than she loved me. I locked away all pens and paper. I locked the door when I left our house and wouldn't let her out.'

'One night, I heard one of my windows break. I leapt out of bed to find that she wasn't there. She had left through my window. I grabbed a torch, went outside, and began searching for her. Finally, got reached the seashore. There she stood, facing the sea.'

'Frantically, I called to her. I told her to come back. I told her that we would find a way for her to be happy.'

'She turned her head to me, smiled, and walked forward into the sea. She kept walking further and further out into the water.'

'I screamed to her. I told her she could not survive the sea.'

'She looked back one last time. She called back to me. She said that it was true that she could not survive in the sea. But, she said, she could also not survive here. At least, she said, she knows the sea. At that, she walked forward. A wave crashed above her head and she disappeared.'

The priest looks up from his ale, his eyes welling up with tears. 'We are all redeemable in the eyes of God', he says. 'You have asked for forgiveness, and it is up to God to grant it to you.'

The Mariner laughs. 'Nonsense Father', he chuckles. 'I need more than your words. I must travel on a path to redemption. I cannot undo what I have done to the one person I loved. However, I can try to give something back to the village I stole her from.'

The priest shakes his head, 'How can you replace a life?' he asks.

'That is why I asked you to meet me at this pub', The Mariner replies. 'Steven, over there, is the owner. He tells me there is a way that I can make a fortune of gold. If I make this fortune, I will bring the gold to the village. When I bring them the gold, they will still likely kill me, but I can die with a cleaner conscience.'

'How will you make this fortune of gold?' asks the priest.

The Mariner finishes the last bit of his ale and smiles. 'Steven has told me about a very special juniper tree. It only exists on one volcano in the far north sea, where most of the ocean is frozen. The tree is unique because it is nourished with the volcanic ash in the soil. It has uniquely adapted to the frigid cold temperatures in the air, and the warmth of the volcanic soil in the ground. There is no other tree like it on the Earth. The fruits of this juniper tree create, what Steven calls, "The Magic Gin." The gin is sought after by people all over the world. They claim it heals the sick. They claim it brings enlightenment. They say it brings eternal youth. While I don't know about all that, I do know that they will pay a hefty amount for one drink. If I can find this tree, and bring the fruits back to Steven, he will pay me in gold. I would bring the gold to the village and change their lives forever. I am a mariner by trade. I have a strong ship that can cut through the ice. I plan to travel the north sea until I find this volcanic island. I plan to search this island until I find the juniper trees. I plan to collect the fruits and bring them back to Steven. I plan to collect the pieces of gold. I plan to bring them to the village on the island. I plan to be redeemed.'

'If you don't want to hear about forgiveness, and you already have this plan, why did you send for me? Why do you need me here?' asks the priest.

'There is more', says The Mariner, 'Of course, I wouldn't normally believe this, but this myth has me uneasy. It is said that the island is guarded by giant whales. Anyone who goes near the island is doomed. The whales will smash a mariner's vessel with their large tails. I need your blessing for my undertaking because I need protection. This is my path to redemption. I set sail tomorrow. I need your blessing tonight.'

At this, the priest stands up, gives The Mariner a blessing, and abruptly walks out of the pub. The Mariner smiles, gives a nod to the owner, and leaves to rest before his journey begins.

If an Environment Should Fail …

The Mariner's sad tale is a lesson in the environment. It demonstrates how our environment is part of us, and how, in turn, we are part of our environment. It shows us that our environment has conditioned us to be a working component of it. This conditioning allows us to survive the specific challenges within it.

The Mariner destroys the woman he loves by removing her from her environment and not allowing her to adapt to her new home. She has skills that have been developed by a lifetime on her island. These skills are all of a sudden irrelevant. She cannot utilise her skills and, as such, lives with boredom. Conversely, her new home presents challenges to her that she has not faced before (new food, cold winters). She hasn't developed the skills to deal with these challenges because they are new and, as such, she lives with the anxiety of having to face these facets of a new environment.

An environment fails an individual when that individual cannot thrive within it. All individuals, of all species, learn skills to survive in the environment they are part of. In this, they are working components of the environment. If an environment is changed to the point where these skills are irrelevant, the individual cannot survive, and the environment fails.

Our environment consists of everything around us and everything external that might influence us. It is the totality of external variables that we live among. At the same time, we are components of it. Our presence, and everything we do, ultimately influences our environment. We depend on our environment and our environment depends on us.

The Mariner's path to redemption is a curious one. He will begin a search for a tree that only exists in one specific and unique environment. Because of the unique environment, the tree is unique and, most importantly, its fruits are unique. They are so unique, in fact, that legends have spread about a magic elixir that can be created by it. The Mariner feels that he has the skills, developed by a lifetime at sea, to seek out the environment where this special tree grows. We are left to wonder if he will be successful.

5.2 Each of Us Is Thrown into a World of External Variables

Our world is determined by what we perceive through our senses. What we feel under our feet, what we feel on our skin, what we see in our vicinity, what we hear in our earshot, what we taste in our mouth, and everything we smell, gets processed through our brains and attached to what we recall. The end result is the world each of us lives in. This is our environment.

All of our sensory perceptions work together to paint a grand picture of the world around us. This is the world we are tasked with surviving through. Each of us are driven, through our genetics, to face the challenge of survival by working through each perception in a quest to feed ourselves and stay safe from danger. Our environment presents us with a collection of challenges in this quest.

Our natural history has granted us genetic skills to manage these challenges. Additionally, the course of our lifetime has granted us learnt skills to manage even more challenges, or have made us more adept at managing greater challenges. These challenges exist as mitigating every external variable one is faced with: acquiring resources, staying safe from potential danger, interacting with conspecifics, raising offspring, finding shelter, or anything else that one might be faced with over the course of any given moment.

An individual's environment is a collection of external variables that define the challenges one encounters moment by moment. If the challenges of one's environment consistently exceed the skills that one has garnered throughout the course of one's lifetime or through the course of the species' natural history, one lives in perpetual anxiety. Conversely, if the challenges are consistently beneath one's skills, that individual lives in a state of perpetual boredom. Luckily, each species has evolved to live between these two extremes – an ecosystem where the skills they possess match the challenges they face.

It is easy to fall into a limited view of the environment. The air one breathes, the terrain one walks on, the sounds one hears, and the visions one sees, are all the *obvious* components of an environment. However, an environment is much larger! One's environment is every external variable that one's senses can perceive. An environment is really every external variable that an individual is faced with and every challenge that an individual contends with.

As care technicians, we are in charge of these variables. This is a big responsibility. We are in charge of, not just the terrain an animal inhabits, but most of most of what an animal will encounter. We are in charge of whether or not to control the climate of an environment – and then how much to control that climate. We're in charge of other con-specifics that an animal might encounter. We are in charge of the colours an animal might see. We're in charge of the sounds an animal might hear. We are in charge of the food an animal eats. We're in charge of the quality of water an animal drinks. We're in charge of the medication an animal receives.

Perhaps most critically, we are tasked with creating an environment where the challenges that are present within it, match the skills that an animal possesses (Figure 5.2). As such, the artificial environment we create must allow for survival in such a manner that it neither creates perpetual anxiety nor perpetual boredom.

Figure 5.2 A frog explores his terrarium at the Nashville Zoo. (Photo credit: author).

A Nocturnal Animal Never Fears the Dark

Because animals are created by the environmental circumstances of their natural histories, they are products of that environment. Because each trait is a tool, granted to that species either through the process of natural selection – allowing an individual to navigate the specificities of their species-typical environment – or of learning – allowing an individual to navigate circumstances that might be unique to their specific population, animal traits are specifically tailored to an environment. As species' skills are tailored to the unique challenges of a unique environment.

When animals are in their natural environment, they live in a state of equilibrium where their skills match the challenges that are present. However, when the environment is changed, or if an animal moves to a new environment, this equilibrium is thrown off. The result can be as significant as an unsurvivable situation for the animal or a state of perpetual anxiety until the animal can develop the skills necessary to navigate their new or changed environment.

For example, for the most part, humans, as a diurnal species, rely on light to get around. As such, humans are, for the most part, a very visual species. Most of the time, our waking environment features daylight. When it is dark, we go to sleep. In fact, darkness can cause disorientation or even fear. When a child's mother turns off the bedroom light and closes the door behind her, the child is left alone in the dark without the ability to see. The child has lost the sensory perception that brings him security. He is left with only his thoughts, which conjure up images of any possible danger that might be lurking within the darkness. As a very visual being, the child doesn't have the skills to manoeuvre through the dark that another person might have – someone with more experience moving around without the aid of vision. The child has even less skills than a species whose other senses allow them to get around in the dark. The lack of the child's experiential skills and genetic skills have created his current state of anxiety.

A nocturnal animal doesn't need to fear the dark because they've been granted the skills to navigate the world with very little reliance on vision through the enhancement of their other senses. These senses have given the nocturnal species the skills to meet the challenges of a dark world.

Environmental Change

In nature, environments are ever changing. Moment by moment, day by day, year by year, and eon by eon, environments fluctuate and change. Some of these changes are cyclical, while other changes are lasting. However, nothing is permanent in an environment. Even lasting change continues to evolve. Each environment has unique features of change. The animals that occupy an environment, being part of the environment, must adapt and change with it. Animals have evolved, alongside their environment, to navigate the specifics of the changes they encounter. Animals adapt to environmental change through adapting body parts, adapting body coverings, or adapting behaviours.[1]

---

[1] See Schluter, 2000.

There are two categories of environmental change: linear change, which is lasting; and circular change, which is cyclical. Within these categories there are countless types of linear and circular changes; and, each can be unique to a specific environment. Animal adaptations that are tailored to such unique changes. These adaptations can prove to be detrimental to an animal if they are removed from their natural environment.

Resiliency is the ability of an animal to adapt to any given change in an environment. Some species achieve resilience through an ability to physically adapt to changes. For example, thermoregulation is an example of resiliency, whereby homeothermic animals can metabolically maintain their internal body temperature regardless of the temperature of their environment. Another example is a species that is able to consume other food sources, outside of their preferred foods; and therefore still eat during times that their preferred foods are unavailable. Species may also be able to achieve resiliency through behavioural adaptations. For example, a species may learn how to evade a new predatory species that is introduced into an environment. Some animals are even able to make and utilise tools to acquire resources, such as chimpanzees.

- *Linear change*

  In **linear change**, an environmental shift persists. What was at first a disturbance, becomes a chronic condition. Linear change can either be a single change that becomes permanent or an ever increasing evolution. Where, at first, an animal will respond to this change with short-term defences, animals will begin to replace defences with long-term adaptive strategies. Over time, these strategies become the hallmark traits of the species.[2]

  Linear change can occur naturally, as in the case of a natural climate shift or a natural shift in the distribution of flora and fauna. Linear change can also occur as **anthropogenic change**, when, for example, humans alter a landscape, or introduce species of plants and animals. Anthropogenic change can occur most significantly in the case of human-induced global climate change. Almost always, anthropogenic changes occur exponentially more rapidly than natural change.

  If linear change is gradual, animals can adapt through both physical adaptations and behavioural adaptations. However, with rapid linear change, animals are more limited in how they can adapt. If the rapid change is significant, animals may not be able to adapt and will, therefore, become extinct in that environment. However, some species are highly adept at responding to rapid linear change through altering their behaviours. This ability is an example of resiliency.

- *Circular change*

  Circular change occurs periodically and regularly. Seasonal variations are an example of circular change. To acclimate to seasonal variations in the environment, animals evolve a plasticity that suits the normal circular change that they would encounter in their natural habitat. The species-specificity of this plasticity is rooted

---

[2] See Young et al., 1989.

in the animal's natural history and is tailored to allow that species to survive in their given environment.[3]

Seasonal changes can affect, among other things, temperature, rainfall, wildfires, and flora production. In some environments, these changes are more significant than in others and, thus, necessitate a higher degree of resiliency in the species that live in these climates. For example, in seasonal wet-dry environments that have massive differences in seasonal precipitation, animals have to adapt to part of the year living in drought conditions and the other part of the year living in flooding conditions.

A key aspect of circular change is seasonal availability or location of preferred food sources. Animals can adapt to this by switching their food preferences according to season. Animals may also migrate to a temporary environment where their preferred food source is still available.

Oftentimes, seasonal food availability is coupled with a significant temperature change. When the temperature drops, animals require more energy to maintain their metabolic rate. This compounds the need for resiliency. Some animals adapt to this by entering into a state of **torpor**. Torpor occurs when an animal goes through a temporary reduction in physiological activity and thus reduces both their metabolic rate and the need for energy rich foods. Some animals may even go into a state of **hibernation**. Hibernation can last for weeks or even months during this time. An animal only requires a very small percentage of the food intake they need during active times.

Another key aspect of circular change is migratory patterns. Animals migrate for a number of reasons. Animals may migrate to search for their preferred food source when those food sources are unavailable within their given seasonal environment. Animals may also migrate to avoid a seasonal presence of a predator. Animals may migrate due to extreme temperature fluctuations and the need to stay within a preferred temperature or climate. Some animals may migrate short distances, while other animals, such as migratory whales, may traverse half the globe in a season. Migratory animals have definitive evolutionary traits that allow them to migrate over these distances.

The Spokes of the Wheel

An environment can be thought of as a wheel with a million spokes. Each spoke plays a small part in how strong that wheel is. Each spoke also plays a small part and how well that wheel will ride. Conversely, the wheel itself dictates how long a spoke needs to be or how thick a spoke needs to be. The presence or absence of each additional spoke determines the dynamics of the wheel. The spokes are part of the wheel, and the wheel does not exist without the spokes.

Likewise, each individual is both affected by their environment, and exists as a part of it. The collective presence or absence of each organism creates the environment.

[3] See Gotthard & Nylin, 1995.

Each individual is part of the environmental makeup. Each individual is part of the environmental change. The actions of each individual determines the way the environment functions. Each individual must adapt to the presence of all the other organisms – conspecific and heterospecific. The unique combination of individuals and their actions are the uniquities of each environment. As products of an environment, individuals are created by our environment. The environment and the individual are mutually tailored to each other. Outside of this environment, welfare is compromised.

5.3   Anatomy of the Environment

The environment is the amalgamation of all external variables that an animal is faced with, and the interactions therein (Figure 5.3). The environment shapes the hallmark traits of a species, as well as the traits of a population and the traits of an individual. Species, populations, and individuals have evolved and adapted to each specific environment. Species, populations, and individuals have been conditioned to their specific environments. Outside of these environments, they either cannot survive or can survive with deeply compromised welfare.

Figure 5.3 Sea lion at the Saint Louis Zoo. (Photo credit: author).

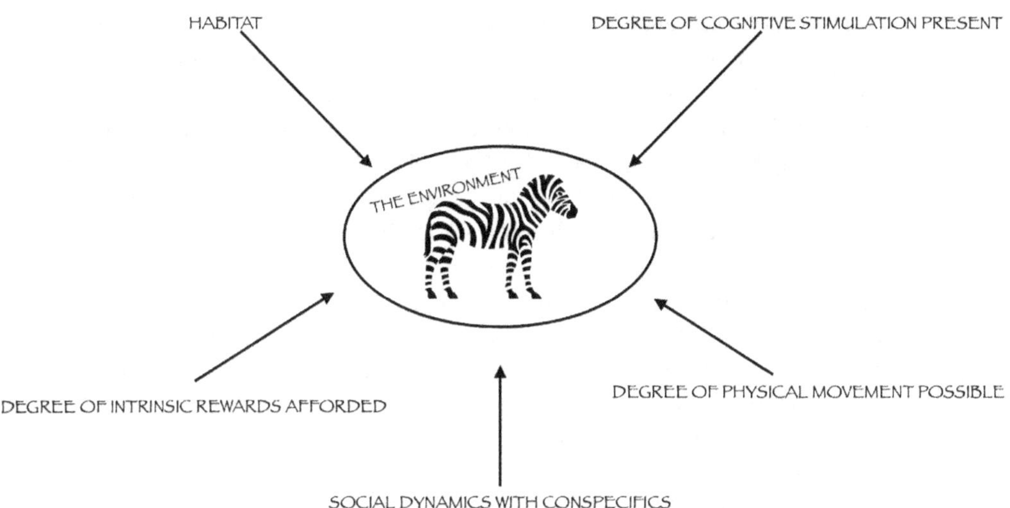

Figure 5.4 The anatomy of the environment.

The traits that animals have developed within their environments can be necessary to their life functions, necessary to their survival, and necessary to their welfare. The inability to utilise these traits is a compromise to all three.

The anatomy of an environment is deeply complex. It includes not just the habitat, but also the degree of physical movement possible, the degree of cognitive stimulation present, the social dynamics with conspecifics, and the degree of intrinsic movements afforded. Each aspect of the habitat helps to define the habitat. Each aspect determines the traits needed to survive within the habitats and determines the adaptive skills that will be developed as a result of living within the habitat (Figure 5.4).

Habitat

Every physical external variable that surrounds you, is your habitat. This includes the terrain you trod upon, the air you breathe, the water you drink, the food you consume, and the presence, absence, and whereabouts of resources. Your habitat is the physical features of your landscape, the temperature of your climate, the weather patterns of your climate, the amount of light, and the amount of darkness. It is what you can feel, what you can see, what you can hear, what you can smell, and what you can taste.

The habitat is the foundation of the environment. All other aspects of the environment hinge on it and its features.

General Aspects of Habitats

There are general aspects of one's habitat that define it. For example: Is the habitat aquatic? Is it terrestrial? Is it mountainous or flat? Is it full of trees or sand? These generalities define the categories of habitats.

The general category of my habitat is going to determine, in the broadest terms, the type of environment in which I can or cannot survive. For example, if my species evolved in an aquatic habitat, I may have developed gills to pull oxygen out of water. I may have developed fins and flukes. I may have developed the ability to hold my breath for long periods of time. I may have developed a body structure that can withstand the high pressures at ocean depths. If my species evolved in a desert habitat, I may have developed the ability to go for long periods of time without water. I may have developed defences from the harsh sun. I may have developed feet that can walk on the soft sand. If my species evolved in a cold mountainous habitat, I may have developed the ability to live at extreme altitudes with the oxygen levels there. I may have developed the ability to withstand extremely cold temperatures. I may have developed a food preference for fatty foods that help keep me warm. All in all, I cannot survive outside of the general qualities of my habitat.

There are obvious examples of animals not being able to survive in different habitats. For example, obviously a whale cannot live in the desert. A tiger cannot live in the ocean. A walrus cannot live in a swamp. However, there are many more subtle aspects of habitats that make some habitats generally survivable and others not survivable at all. For example, some animals might require a specific type of microclimate within a habitat. **Microclimates** are small areas of a perpetual climate variance within a larger habitat. A mountain casting a shady and casting a shadow over an area of valley would create a microclimate. A waterfall emptying into a stream creates a certain microclimate within the stream. A warm spring in a cold environment creates a microclimate. Microclimates can even be created by human structures. A power plant creating a warm area along a river creates a microclimate. The shadow of a bridge creates a microclimate. Some species have evolved to live specifically in these types of microclimate. Without them, the habitat is not survivable for them.

Unique Aspects of Habitats

There are also unique aspects of one's habitats that can be just as significant to populations or individuals. For example, it is specific to one's habitat where a stream might be located. It is also specific to one's particular habitat where a preferred food source might be located. Where potential dangers lie are unique to specific habitats; as are where areas of safety lie.

Whereas, animals tend to adapt on a species level to the general aspects of their habitat, animals adapt at the population level, or even individual level, to the unique aspects of their habitat. Oftentimes the unique aspects are adapted to behaviourally, while the more general aspects are adapted through genetics. Though the general aspects of a habitat may define the hallmark traits of a species, the need to adapt to the uniquities of a specific habitat may be no less significant. In fact, population-level and individual-level adaptations are often critical for survival; and as such, a population or individual may not be able to survive outside of that specific habitat.

Degree of Physical Movement Possible

How much an animal needs to move, and how much an animal *can* move are a defining features of an environment. Over time, species have adapted to the degree of physical movement that their environment affords them. Additionally, they must adapt to the degree of physical movement that is required in an environment.

Obviously, migratory animals are going to have a great deal of physical movement within their habitat; which, in some cases, can stretch the greater part of the Earth. Nomadic animals (which may have very large home ranges) are also going to have a habitat that affords them a very large degree of physical movement. Conversely, burrowing animals are going to have much more restricted space, and therefore, have much less ability for physical movement within their habitats.

Locomotor Movement

Locomotor movements are travelling movements. Movements that allow an animal to traverse from one part of their habitat to another part of their habitat, are locomotor movements. The size of the habitat is going to dictate the amount of locomotor movements afforded to the animal. Some species of whales travel over 200 kilometres per day. Chimpanzees can cover over 10 kilometres in a day. When a blue jay is migrating, they can travel over 800 kilometres per day. These are obviously species with a great deal of locomotor movement afforded within their environments. However, a chipmunk, on the other hand, may have an entire home range of only a quarter of a hectare.

There are several determining factors in how far an animal can locomote within their environment. Chief among these are dietary needs. Omnivores can obviously afford less of a home range than carnivores (who have to hunt for prey that might also have large home ranges). Animals that have to migrate seasonally to find their preferred food resources obviously have to travel very far distances for this to occur.

Many of the hallmark traits of a species are based on the type and degree of locomotor movements that are required and afforded by an environment. Limb structure, musculature, endurance, and even metabolism, can be determined by it. If animals are not afforded the degree of locomotor movements that normally exist within their natural habitats, it can significantly, and even sometimes gravely, affect their welfare.

Non-locomotor Movement

Movements that are intended for any other reason besides locomotion are **non-locomotor movements**. Such movements may be **manipulative movements**, whereby an animal is manipulating an object or part of their environment for any given reason. These also may be **communicative movements**, movements that are intended for gestural or visual communication.

These movements could be non-communicative **socially interactive movements**, which would include fighting, grooming, parenting, breeding, or any other socially interactive movements between conspecifics. Other non-locomotor movements are **self-maintenance movements**. These are movements that are intended to satiate a personal need. Examples of self-maintenance movements include scratching, self-grooming, stretching one's limbs, and shaking off water.

There are an array of other non-locomotor movements. Many of these movements are species-specific. Their existence in a species is dictated by the environment where they evolved, based on what they contend with in their habitats.

Body Size to Space Ratio

Body size is a determining factor in the degree of movement that an individual has. A large fish in the same pond as a small fish, has a much less degree of movement simply because he requires more space. An ant in a large and complex mound has a great deal of physical movement, even though that mound might seem very small to a gorilla. Therefore, in many cases, the size of a habitat that one needs to survive, is going to be dictated by body size.

Degree of Cognitive Stimulation Present

Whether a species possesses a complex brain, or a simple central nervous system, they survive by a mental process. This process can include the following at varying degrees (depending on the species):

- A spatial map of the relevant habitat;
- A memory bank of past experiences that can be recalled;
- An identification guide of frequently familiar conspecifics;
- Where resources are located and how to acquire them;
- A knowledge of how to avoid or get away from danger;
- A knowledge of parents, offspring, or other relevant relationships;
- A bank of skills necessary to survive one's environment.

A mental process includes all information entering through the senses (experience), combining this with a perception of the information (through genetics) and responding with a reaction based on the perception (experience and genetics). Mental processes are at the foundation of how an individual survives an environment.

Mental processes represent a marriage of what is acquired through genetics and what is acquired through learning and experience. Sometimes experience and genetics are thought of as distinct binaries, but they actually are not. The ability to learn (and the drive to learn) is, after all, a genetic ability. An animal's natural environment determines what mental processes are necessary, as well as what cognitive processes are available to be utilised. As such, all animals are genetically predisposed towards exercising a species-specific mental process. Without this process, or the ability to exercise it, the animal will not survive. Mental processes are as critical as physical processes, because they drive the physical process. They determine how animals will react to danger, the presence of a resource, or to each other.

Each mental process builds on a previous process. **Conditioning** occurs when animals store a mental process and reaction to a given stimuli or scenario. Animals can become conditioned to certain stimuli; and, in turn, create and maintain a conditioned response. In this way, animals are storing the experience of a mental process and utilising it in future similar scenarios. This is another tool for survival.

We can use a bigger term to define all of this – cognition. **Cognition** includes: the mental process, the perception and awareness that is created, and the conditioning that comes from it all. Cognition is a genetic trait that is enhanced through learning. **Cognitive stimulation** occurs each time a cognitive process is engaged. Cognitive stimulation affords the animal more skills to survive. Cognitive stimulation is integral for survival.

The degree of cognition required to survive an environment exists as the degree of cognitive stimulation present within the environment. Animals adapt to this degree of cognitive stimulation and require it for survival and welfare.

Social Dynamics with Conspecifics

The social dynamics of a population are key components of the environment. Population size, sex ratio, and even social roles, are major factors of how individuals survive within their environment. Social dynamics determine how individuals are reared. They factor in the way individuals stay safe. They play a role in how individuals acquire resources. They aid in how individuals learn the behaviours necessary to survive in their specific environments. They dictate how a species reproduces.

Social dynamics are ultimately determined by a habitat. The amount and distribution of available food resources, along with the amount of cooperation or competition required to obtain them, are major factors in determining group size and group dynamics. For example, a large animal whose preferred food source is sparsely located around a habitat, may only be able to afford a very limited social group. In some cases, this species may even need to live a solitary lifestyle. Conversely, species that live with an abundance of resources, with very little competition for them, can generally afford a much larger social group.

Additionally, a species metabolic rate, combined with the chemical energy within the food source itself, is going to be a factor in social dynamics. High-energy foods that can satiate the given species for a long period of time, can lessen the amount of competition for resources. This can even be true when food is sparsely located. This can allow a species to afford larger social groups.

Reproductive selection is also going to play a part in social dynamics. *K*-selected **species** live in relatively stable environments. As a result, they have single-offspring births. *R*-selected species live in unstable environments. As a result, they have multiple offspring births (giving them a greater chance of reproductive success). *K*-selected species are obviously going to have a much different social dynamic and different group composition than *R*-selected species. This can affect group size, parental investment, reproductive strategies, among other variables.

These social dynamics, required by a species through the course of their evolution, are integral to their survival and welfare. Deprived of these dynamics, the individual faces an instinctual stress of having to navigate a world without what they need for survival. For example, if a solitary animal were to find themselves being forced to live among conspecifics, they would face undue competition, potentially grave aggression, and the anxiety inherent in such a situation. Likewise, a social individual finding themselves in a solitary environment where they are without the aid and companionship of conspecifics, faces an environment where they may lack the protection they have evolved to be accustomed to, may lack the cooperation they require, or may lack the means of social transmission and learning necessary to survive.

Degree of Intrinsic Rewards Afforded

An **extrinsic motivation** is an action that results in an immediate and clear ecological reward, such as eating, breeding, or fleeing danger. For example, being driven to acquire a preferred food source is an extrinsic motivation. Its reward is consuming the preferred food source and the satiation that it provides.

On the other hand, an **intrinsic motivation** is the motivation to perform an action for the sake of doing it with no clear or immediate reward. An **intrinsic reward** is the satisfaction of performing an intrinsic action. Humans spend a great deal of time engaging in intrinsically motivated actions. Playing sports, playing games, watching films, watching television shows, listening to music, exploring new areas, learning new things, engaging in hobbies, and many other things that humans participate in on a day-to-day basis, are examples of intrinsic motivations that carry intrinsic rewards. Playing a football game has no clear or immediate reward. However, the pleasure of playing the game is the reward itself.

Likewise, non-human animals also have intrinsic motivations. These motivations are present within their natural environment, and the degree of the rewards they offer are something that this species has grown accustomed to. We make a mistake when we think of intrinsic motivators as strictly in the domain of humans (or even just in the domain of the higher primates). Intrinsic motivations abound in the animal kingdom. Moments of play, moments of exploration, moments of performing a task when the reward can be achieved far easier. The animal kingdom is rife with these examples.

The degree of intrinsic rewards afforded by an environment depends on all other factors in the environment: social dynamics, degree of physical movement possible, degree of cognitive stimulation present, and the habitat. As with other factors of the environment, animals adapt to the degree of intrinsic rewards that are present in an environment, as a species, over the course of their natural history.

The following are places where animals frequently find intrinsic rewards within their environments:

- *Intrinsic exploration*

    Intrinsic exploration involves an animal exploring unfamiliar aspects of their habitat with no clear reward for doing so (i.e., there is no perceptible preferred food source in the unfamiliar area, there are no potential mates in the unfamiliar area, etc.). The presence or amount of species-typical intrinsic exploration varies greatly between species. Sensory change, itself, may be a motivator for intrinsic exploration.[4]

- *Play*

  Play has been observed in an array of animal species (most frequently in mammals and birds). It has long been used as an indicator of animal welfare due to the appearance of it being accompanied by an observable pleasurable emotional state.[5] The social dynamics of the species is obviously going to be the biggest driver in the presence and frequency of play.

- *Contrafreeloading*

  When animals perform seemingly skilled tasks to obtain a resource, when the resource is freely available without performing this task, this is known as contrafreeloading.[6] For example, a woodpecker may continue to drill through the trunk of a tree to obtain larvae or insects, while the same food resource might exist outside the trunk, and be obtainable without drilling. Contrafreeloading is an example of an action that appears to exist for its own intrinsic reward.

5.4 The Natural Environment versus the Artificial Environment

In order to compare an artificial environment with a natural environment, we must first define what an artificial environment is. An **artificial environment** is any animal environment where any aspects of its anatomy have been replaced and replicated by human agency. The most obvious example is a typical captive environment, where an animal's habitat has been replaced by an artificial habitat. However, this is not the only way an artificial environment can exist. For example, if an animal's movements are constrained in any way, even in their natural habitat, an artificial environment is created. If the social dynamics of an environment have been compromised by human agency, an artificial environment is created. In fact, there is an array of ways that humans can transform a natural environment of an animal into an artificial environment.

In Chapter 6, we will explore how an artificial environment might compensate for some aspects of a natural environment. However, for now, let's examine what an artificial environment lacks by its very definition.

Flow

As we have explored earlier, the interactions of a species with their environment is an interrelationship between the challenges implicit in the environment, and the skills that the species possesses to navigate these challenges. The species have developed

---

[4] See Hughes, 1997.
[5] See Held S. D. & Špinka, 2011.
[6] See Osborne, 1977.

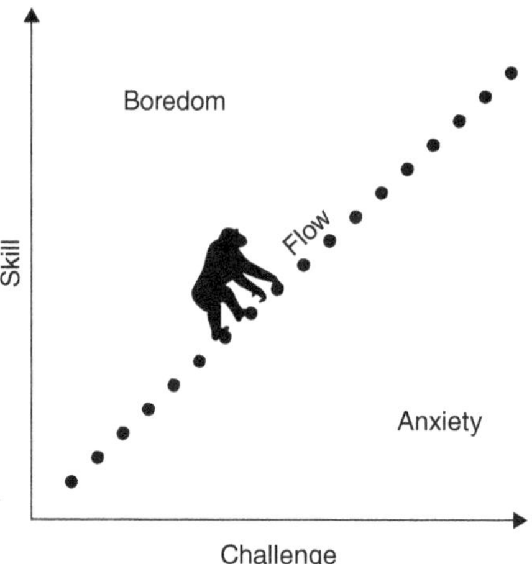

Figure 5.5 Flow is the balance of challenges versus skills for each individual in an environment.

these skills over the course of their natural history. An individual is born with these skills, through genetics, and refines these skills over the course of a lifetime.

If the challenges present in the environment exceed the skill of the individual, that individual either fails to survive or lives in a state of anxiety. However, if the individual's skills outweigh the challenges of the environment, the individual lives in a state of boredom. Luckily, a species has evolved within their natural environment and, as such, lives within an equilibrium where the individual's skills match the challenges within. This ecological equilibrium is known as flow (Hintze & Ye, 2023). **Flow** is the balance between the appropriate level of skill that a species possesses to meet the specific challenges of an environment (Figure 5.5).

An artificial environment has the disadvantage of not developing this equilibrium. Because there has been no course of natural history, there is no natural flow in a created environment. Because of this, the challenges in the artificial environment may be insurmountable to the level of skill that the individual possesses. Therefore, the individual either may not survive this environment or lives in that state of perpetual anxiety. Conversely, the artificial environment may lack the challenges necessary for the individual to utilise their natural skills meant for their natural environment. In these cases, the animal will experience boredom. Thus, one of the primary deficiencies of an artificial environment is the lack of flow within. Without efforts to compensate for flow, the individual's welfare is deeply compromised and thriving is impossible in an artificial environment.

Standards and Uniquities

Deep below the standard dynamics of an environment, there lie the uniquities. The big standards are easy to replicate: for example, aquatic saltwater habitats, aquatic freshwater habitats, desert terrain, swamp terrain, and rainforest terrain. However, just as important as these standards are the environmental uniquities. In many cases, these uniquities are all but impossible to replicate artificially. As such, without a novel approach to creating and maintaining an artificial environment, it can indeed become unsurvivable for some species.

- *Seasonality*

     Animals adapt to the seasonality of weather patterns, resource availability, temperature variation, precipitation variation, and anything else that might seasonally affect a species. In some cases, the adaptations to seasonality are significant hallmark traits of a species – dictating feeding patterns, migratory patterns, times of torpor and hibernation, and other key aspects of the survival of a species. A fixed artificial habitat, without seasonal variation, may not only be monotonous for an individual, but may in fact not be survivable if those adaptations are not properly compensated.

- *Microclimates*

     Many species adapt to the presence of microclimates within their habitats. Microclimates may be necessary for temperature regulation. Microclimates can also be factors for what resources are available. Microclimates may offer shelter from weather conditions. Microclimates may even provide safety from predators. If artificial environments do not replicate microclimates, it can lead to anxiety or unsurvivable conditions for the animals within.

- *Biodiversity*

     The types of flora and fauna within an animal's environment is at the foundation of the adaptations they possess; and, therefore, at the foundation of the hallmark traits of the species. It can be very difficult to replicate the biodiversity found in a natural habitat within an artificial habitat. Without the appropriate biodiversity, an animal's skills may no longer match the challenges that are present. This makes flow impossible and can even make the conditions unsurvivable. The amount of available resources, the ratio of predator to prey, and the ability to utilise habitat structures, are all deeply compromised when the biodiversity is not replicated.

Cognitive Stimulation

As we have explored, the mental processes of a species are specifically tailored for survival in a given environment. When these mental processes are exercised and utilised, cognitive stimulation occurs. Animals adapt to the type, degree, and amount of cognitive stimulation present within the habitat they have evolved within.

Cognitive stimulation is a key component of flow. Utilising one's skills to navigate the challenges within an environment grants one the cognitive stimulation required for welfare and survival. Since animals within their natural habitat live within that equilibrium of flow, they have evolved to experience a specific degree of cognitive stimulation.

An artificial environment may not provide this stimulation. When this occurs, animals fail to utilise the cognitive skills that they have evolved to use. Boredom and cognitive lethargy are the results.

Social Dynamics

We have explored how the social dynamics within an environment are critical to the survival and welfare of every individual within it. The continued existence of a population within an environment hinges on social dynamics. Reproduction, staying safe, and acquiring resources all have social components.

Artificial environments lose many of the components of natural social dynamics. Natural fluctuations in the population, a natural sex ratio, a natural group size, emigration and immigration, and solitary versus social lifestyles, can all be compromised in an artificial environment. Without novel methods of replicating all aspects of a species's natural social dynamics, animals in artificial environments can suffer from anxiety, undue competition, undue aggression, and even death.

Space

The space to move, the habitat to traverse, and the migratory patterns to follow, are all aspects of an animal's strategy to survive. The strategy is implicit in their ecologies and must be followed for an animal to adequately navigate the challenges that are present within their environments. Additionally, the means by which an animal locomotes, the way an animal utilises their limbs, and the way an animal moves their body, are part of this process. If this movement is confined, as it so often is in artificial habitats, an animal's welfare can be deeply compromised.

It is difficult to imagine an artificial habitat being sufficiently large enough to adequately house a migratory species. It is also difficult to imagine an artificial habitat replicating the terrain of a novel habitat (such as a glacier or a tundra forest). It is even more difficult to imagine an artificial habitat being adequately able to house an animal that lives within the deepest parts of the ocean. However, animals that live within these habitats are some of the animals that find themselves housed within artificial environments. As such, it is crucial for those that control artificial environments to find novel means of replicating the amount of space, and the type of space, that an animal has been deprived of in captivity.

Freedom of Choice and Self-Determination

The freedom to explore (where and when one wants to), the freedom to choose what to eat, the freedom to choose when to eat, the freedom to socialise with chosen conspecifics, the freedom to sleep where one wants to, the freedom to awaken at a desired time of day, and the freedom to perform intrinsic actions, are often taken for granted. Within an artificial environment, these freedoms may not exist. Space is limited, schedules are managed, conspecifics are fixed, and food is offered without choice.

While these are often necessary factors within an artificial environment, they can deeply compromise an animal's welfare. For such an environment to foster positive well-being, avenues of freedom of choice and self-determination must be available for each individual living within it.

Implications

Every individual, at birth, is thrown into the abyss of existence. Luckily, their genetics have adapted to the specific environment where their species lives, and have given them skills to survive it – providing them, in a manner of speaking, a soft place to land. Through experiential learning in this environment, the individual develops even deeper skills. All individuals (of any species) are experts at living within their environment. They live as individual parts of it.

The anatomical structure of an environment contains a vast array of components. Many of these components are not easily apparent. However, though inconspicuous, these components can be just as significant as the most obvious aspects of an environment.

An artificial environment, in order to achieve the survivability of a given species that it houses, attempts to successfully replicate a natural environment. However, it can be very difficult, if not impossible in some cases, to completely replicate many of the components of a natural environment. This can result in a species that lives in an abject state of poor welfare. In many cases, this state of welfare is not survivable.

In Chapter 6, we'll explore how artificial environments can attempt to compensate, if not replicate, the components of a natural environment. This is one of many facets of the complex science that is animal care.

References

Gotthard, K. and Nylin, S., 1995. Adaptive plasticity and plasticity as an adaptation: a selective review of plasticity in animal morphology and life history. *Oikos*, 74(1), pp. 3–17.

Held, S. D. and Špinka, M., 2011. Animal play and animal welfare. *Animal Behaviour*, *81*(5), pp. 891–899.

Hintze, S. and Yee, J. R., 2023. Animals in flow – towards the scientific study of intrinsic reward in animals. *Biological Reviews*, *98*(3), pp. 792–806.

Hughes, R. N., 1997. Intrinsic exploration in animals: motives and measurement. *Behavioural Processes*, *41*(3), pp. 213–226.

Osborne, S. R., 1977. The free food (contrafreeloading) phenomenon: a review and analysis. *Animal Learning & Behavior*, *5*(3), pp. 221–235.

Schluter, D., 2000. *The ecology of adaptive radiation*. Oxford: Oxford University Press.

Young, B. A., Walker, B., Dixon, A. E. and Walker, V. A., 1989. Physiological adaptation to the environment. *Journal of Animal Science*, *67*(9), pp. 2426–2432.

6 Compensation in an Artificial Environment

Figure 6.1 A chimpanzee enjoys enrichment at the Kansas City Zoo. (Photo credit: author).

6.1 The Iceman's Tale

In a place so far to the north that no one knows it exists, the world is white. The snow on the ground is white. The sky is white. The icy ocean is white. In fact, everything is so bland and colourless that one's eyesight is completely useless. Here, the cold world appears lifeless. The sun never peaks out from the grey clouds. The stars never shine at night.

Deep in this frosty world, a man lives alone. It is unclear how he got here. It is even less clear how he survives. In fact, he has no memory of his origins. As for survival, he only knows that he is alive so something he is doing seems to be working. He knows nothing else but this world. He is The Iceman.

The Iceman begins his day like every other day – hiking through the snow. He is clad in an intricately woven cloth of unknown origin, layered with skins of various animals of unknown species. In one hand, he holds a smooth wooden spear. On the other hand, he holds a smooth black stone.

In the distance, where the ice gives way to ocean waves, a large whale tail emerges from the water. The tail splashes against the ocean surface, creating a spray of water and a large wave that runs over the ice. The tail disappears under the water. The sounds of a very low rumble emits from the ocean and travels over the ice. The sound reaches The Iceman. He looks in the direction of the sound and shudders with fear. He dares not approach the ocean. The whales scare him.

He shuffles through the icy ground until he hits a spot of smooth ice. Kneeling down, he lifts the black stone and smashes it against the ice. He repeats this again and again until the ice begins to crack, revealing a dark body of water underneath. As soon as there are enough cracks in the ice, he plunges the smooth spear deep into the ice. Water splashes out from the ice and freezes as it spills on to the ground. He moves the spear around as if he is searching for something. When the spear hits something, he lifts it up a bit, and plunges it, with great strength, back down. With all of his might, he lifts the spear back up, through the ice. At the end of his spear is an odd-shaped crustacean. The crustacean withers briefly before the last remnants of his life drain away. The Iceman pulls the crustacean off the spear and shoves it into a deep pocket.

He begins to hunt through the ice for another crustacean. However, he hears another low rumble blowing from the direction of the shore. The whales are out. He looks around, then instinctually walks quickly back in the direction he came from.

As The Iceman plods along, following his previous footprints, an outcropping of black rocks appears in the distance. The site interrupts the endless white. As he draws nearer to it, the outcropping becomes larger and obscures the snow and grey sky. Appearing behind it, through the haze, is the outline of a large mountain in the distance. The mountain has a hint of red glow, smoke billows from the top. The world, which just a short distance away was lifeless and bland, is transfixed in this space.

The Iceman stands at the foot of the rock formation. The black rocks tower so far over his head that they appear to scrape the grey sky above. The Iceman approaches an entry point to the interior of the rocks. As he enters, he descends into a black cavern.

The lower he descends, the more his world changes. Finally, he enters a large room. Several large fire pits, with chimneys piercing the rocks above, illuminate and warm the room.

Tapestries adorn the wall. Some are painted with different landscapes (real and imaginary). Some are painted with just splashes of colour. Some are blank. Beside one of the blank tapestries is a small table with four small jars and a brush.

The cave is also furnished with hammocks, chairs, and more tables. On one of the tables rests three handmade flutes. The rugged instruments are made out of an unknown type of wood. Beside the flutes is a drum made out of shells and animal skins.

Resting atop one of the fire pits is a cauldron of boiling water. The Iceman walks over to the cauldron, takes the dead crustacean out of his pocket, and tosses it into the boiling water. Feeling warm, he removes his heavier garments and throws it across one of the chairs.

As The Iceman waits for his meal to cook, he approaches one of the blank tapestries (the one next to the table with the jars). He stares at the black canvas for a bit and then opens one of the jars. He gently dips the brush into the jar and begins to paint. The painting he begins will soon show a world of colour, a world of unique landscapes, a world teeming with life, and a world that he imagines.

He becomes so immersed in his painting that he momentarily forgets about the meal cooking in the cauldron. Quickly, he runs over to it, grabs a pair of homemade tongs, and pulls the fully cooked crustacean out from the water and throws it on a table. As the crustacean cools on the table, he walks to a basket in the corner of the room and pulls out a handful of juniper berries. He brings them over to the table and then pulls up a chair. Taking his time, he eats his meal of crustacean and juniper berries.

When he finishes his meal, he looks back over to his painting in progress. He walks over, opens up another jar, and sees that it is empty. He looks down at his brush and then up again at his painting. He smirks.

He walks over to the table with the instruments on it. He picks up one of the flutes. It is cracked. The sound it makes is wispy and without much tone. He puts the flute down. With a determined look, he grabs his heavy garment from the chair, puts it back on, and leaves the cave.

Once outside The Iceman looks around at the white and lifeless world that surrounds him. He begins to walk. Only this time he walks in the direction of the looming mountain in the distance. The closer he gets to the mountain, the more distinct its shape appears. The faint reddish tint becomes an obvious flow of lava that feeds other large black rock outcroppings all over the landscape. The heat from the volcano warms the landscape from its otherwise icy condition.

The Iceman spots a certain group of rocks and walks towards them. As he approaches, he finds a crevice that he can fit through. Once he squeezes himself through and reaches the other side, his world is transfixed yet again.

In front of him is a large valley of evergreen juniper trees. The landscape is surrounded by a line of black rocks. The trees are healthy and thriving, springing up through the black soil and reaching high up into the sky. The Iceman walks to one particularly large tree. He looks through the branches and finds four particular berries on the branches.

Each berry has a different quality. One of the berries is very dark. He immediately pops this one in his mouth, chews it up, and swallows it. Another berry is deep red. He squeezes it. Out of it comes a thick red dye. The next berry is yellow. When he squeezes this, a thick amber dye emerges. Finally, the last berry is a dark blue. When he squeezes this one, dye that is the colour of the sea pours out. He smiles and places them all in his pocket.

He walks further through the evergreen forest, picking up branches that have fallen off the trees. Some of this wood can be used to make more instruments. Some of the wood can be used to carve into spears, furniture, or tools. The softest of the wood can be used to sew together and actually create a garment.

Feeling satisfied with his journey, The Iceman organises his findings and leaves the magic forest. He plods through the snow to return to his home – deep inside the black rocks.

Separation

The Iceman's Tale is that of an individual who has, either by circumstance of birth or circumstance of event, lives outside of his natural environment. As such, he is *separated* from his natural environment. He lives in a world that is not conducive for human survival. Yet, he survives. He has done so by creating for himself an artificial environment. He has some, but not all, of the factors that allow him to both survive and have some measure of being able to thrive. However, there are still some significant barriers to his well-being. Some key points:

- *The Iceman's origins are unknown.*
  As with many animals we may care for in an artificial environment, we do not know how The Iceman ended up in this scenario. Perhaps he was born in the ice. Perhaps he was shipwrecked long ago. Either way, his condition is that of being separated from his natural environment – an environment where, through the course of his natural history as a human, he is inclined to thrive. Thus, there must be methods of compensation for this in his current environment.
- *The Iceman has created an artificial environment for himself, which allows him some measure of thriving.*
  The Iceman has created an environment where he has both the means of survival (warmth, shelter, food, etc.) and some means of thriving (finding intrinsic reward in painting, mastering musical instruments, exploring, etc.).
- *There are aspects of The Iceman's environment that do not compensate for what he is lacking.*
  Despite an artificial environment that allows him a survivable habitat, freedom of movement, cognitive stimulation, and even intrinsic reward, The Iceman lives without any socialisation at all. We can imagine that his well-being suffers from such a condition.
- *This environment is not natural for The Iceman. He deals with native wildlife. He deals with a challenging climate.*
  The Iceman's fear of the whales is indicative of living among conditions that are unnatural for him. His artificial environment offers him safety from the perceived danger of the whales. His behaviour allows him to avoid the areas where the whales are located. Thus, he does not enter this part of his habitat.

When we care for animals in a created, or artificial, environment, we must compensate for the effects of separation from an environment where they are naturally adapted to living and thriving. Creating and maintaining this environment is our primary role as care technicians.

6.2   The Artificial Environment

The anatomy of any environment, whether natural or artificial, is the same. There is always a habitat. There is always a degree of physical movement possible. There's always a degree of cognitive stimulation present. There are always social dynamics with conspecifics (even if there are no other conspecifics, a solitary environment is the social dynamic). There are always a degree of intrinsic rewards afforded (Figure 6.2).

In a natural environment, animals have adapted specifically to the very unique variables of the environment. They have adapted to millions of years of natural history. In an artificial environment, an animal has been removed from their natural environment (whether by birth or relocation) and placed in a situation that attempts to replicate these natural variables as best as possible in order to meet the species-specific needs of the individual. When these needs are met, an animal can survive. When they are not met, an animal either cannot survive, or survives with deeply compromised welfare. However, as we have seen in the throughout this book, just surviving should never be a benchmark for a care technician in charge of the care and well-being of an animal. Thriving should, in fact, be the benchmark.

Thriving, of course, means the ability to pursue living life to the fullest. Thus, thriving requires replicating not just the variables of a natural environment that prolong life functions. Rather, the ability for thriving requires an artificial environment to replicate the aspects of a natural environment that allow the ability to live life to the fullest. This is a large and complicated process that requires a deep understanding of a species, its ecology, and its natural history. This understanding is necessary for habitat design, the design of available space, the design of a space that enables natural movement, the design of an environment that promotes cognitive stimulation, the design of an environment that contains the possibility for intrinsic reward, and the strategy for what degree of social housing should be afforded.

The Unknown, Unforeseen, and Unavoidable

The amalgamation of every external variable that an individual is faced with in their natural environment includes both the very obvious variables and the inconspicuous.

Figure 6.2 Cotton top tamarin at the Chattanooga Zoo. (Photo credit: author).

The very obvious are those that support life functions, or are the large physical attributes of a habitat, or the highly observable features of an environment. There are also, however, the much less obvious (and even hidden) variables. Animals have adapted to these as much as the obvious ones. This may include everything from the microbiota of an environment to less obvious geographic features, or even the presence or absence of migratory species that only temporarily occupy environments. However seemingly insignificant, species have in fact adapted to these variables, possibly just as significantly, as the obvious features.

In an artificial environment, these hidden or inconspicuous elements may be missed when attempting to replicate a natural environment. In some cases, the absence of these elements may make an artificial environment unsurvivable for a species (or at least 'unthrivable'). Similarly, artificial environments *themselves* may have factors that are less obvious or inconspicuous. Though they may seem insignificant, their presence in an artificial environment may cause it to be not conducive to good welfare for the animals within.

- *The Unknown*

   There may be unknown factors that may have significant repercussions for the animals. For example, the microbiota of an artificial environment may be so different from what an animal has adapted to, that their welfare can be compromised. Also, an artificial environment may be shared by other species (such as smaller animals including insects) that may make the environment inhabitable for certain species.

   There are also the effects of what's going on immediately outside the artificial environment that may have a significant effect on the animals. When occurrences outside the environment begin to affect what is going on inside the environment, these outside occurrences in fact become part of the environment. Such is the case with the presence of observers or visitors. Such is the case with sound pollution or light pollution. Such is the case with neighbouring artificial environments that may be able to be sensed by the resident animals (through scent, sound, or being visible).

   o *The Gut Microbiome*

   Gut microbiota are microorganisms that reside in the digestive tract of all animals. The existence of various make-up of these microorganisms determines all manner of an animal's physical, and even behavioural, ecology; allowing for digestion, dietary choices, resistance against pathogens, immune function, and a host of other life functions. The gut microbiome is determined by the microbiota of an environment. Thus, when an environment changes, the gut microbiome changes. Animals must then adapt accordingly to this change. Like all other ecological changes, this adaptation can take generations of evolution.

   In an artificial environment, the gut microbiome is going to be different than that of an animal's natural habitat. Scientists work to study the effects of the change in microbiome on animal's in artificial environments and find novel methods to attempt to replicate the natural microbiome, such as close housing, or mixed housing, with other species, whose presence can create a similar microbiome.[1] However, oftentimes this can be impossible to replicate, and even difficult to know its effects. For this reason, we much consider the gut microbiome of an artificial environment as a key 'unknown' variable.

- *The Unforeseen*

  There also may be unforeseen factors of an artificial environment. These factors may include immigrating local fauna, the growth of toxic flora, or even emergency weather events. These factors can also include extremely random circumstances that can occur, such as maintenance issues, social issues between conspecifics, and random behavioural or medical issues. These can all have profound effects on the ability to thrive for an animal in an artificial environment.

- *The Unavoidable*

  And finally, there are the unavoidable factors of an artificial environment. Things such as local climate, local air quality, and the logistic constraints of space or social environment are all going to have a profound impact on an animal's ability to thrive in an artificial environment.

When the above-mentioned factors cannot be mitigated, animals should not be held in this artificial environment. However, in many cases, these factors can be mitigated and can be made to have less significance on an animal's ability to thrive.

6.3   The Artificial Habitat

The habitat, or the physical external world that an individual occupies, can best be looked at in terms of function. Animals employ their habitats with a functional purpose. The function may be critical to survival or, at the very least, it may be critical to the ability to thrive. Thus, without it, an animal's welfare can be deeply compromised.

As with everything else in an animal's environment, animals have adapted to the various functional qualities of their environment over millions of years of natural history. These adaptations, and the interactions an animal has with their habitat, are key foundations of a species' overall ecology. Additionally, over the course of their natural history, bodily structure, behaviours, and life functions, have all evolved to conform to the various elements of a habitat and the functional qualities relevant to that species.

Function in an Artificial Habitat

The effectiveness of an artificial habitat can be determined by examining its **functional equivalency**, or the degree in which an artificial habitat replicates, or synthesises, a functional aspect of a natural habitat relative to the particular species. For

---

[1] See Diaz & Reese, 2021.

example, an animal that has evolved by living on soft terrain has adapted *functionally* to it – locomoting across soft terrain, being comfortable on soft terrain. If this animal is deprived of the terrain it has adapted to, this functional relationship is no longer relevant. In this case, the adaptation of interacting with soft terrain is replaced by having to interact with hard terrain. The adaptation to living on soft terrain no longer has any function – and, in fact, may now be maladaptive. Conversely, animals that have evolved living upon hard terrain are going to have a functional issue if they have been moved onto a soft terrain habitat. This can affect everything from the ability to properly locomote to general comfort.

Functional equivalency (or lack of) can be easily observed. Gibbons are small arboreal apes that live in the dense rainforests of Southeast Asia. They have a complex natural habitat that can be difficult to replicate artificially. In the wild, gibbons rarely come to the ground. They do almost everything in the trees. In fact, gibbons have an ecological taboo against going to the ground due to the presence of terrestrial predators. Therefore, gibbons find everything they need in the treetops. They find their food in the trees. They find their shelter in the trees. Their method of locomotion (called brachiating) allows them to move from tree to tree, extremely rapidly. They have evolved long upper limbs and short lower limbs that allow them to have this type of movement. Through brachiation, they are able to get through the thick forest canopy in a completely arboreal manner.

Obviously, a type of habitat where gibbons can brachiate and live an almost entirely arboreal existence can be challenging to create. In some cases, there may be very little ability to brachiate in an artificial habitat. In these cases, one might observe gibbons spending an inordinate amount of time on the ground – which is not a natural proclivity for them. In fact, it is anathema both behaviourally and physically to them. Because gibbons are structured with very long arms and very short legs, their ability to move about terrestrially is compromised. Gibbons' arms are so long, and their legs so short, that they have to walk bipedally while holding their arms over their head. Not only is this unnatural method of locomotion inefficient, but it is also uncomfortable. This can lead to both physiological problems, and even psychological problems such as anxiety due to being on the ground more than they are naturally inclined to be.

However, novel habitat design has devised a solution. In some zoos, gibbons have been given brachiating structures to allow them to move around arboreally. The presence of these structures, combined arboreal shade and aboral surface areas, allow gibbons to live closer to how they would live in a natural habitat. In doing this, these facilities have achieved a functional equivalency.

Naturalistic?

When I was young, I visited the Atlanta Zoo. It was then that I first came to know an incredible individual: a gorilla named Willie B. When I first saw Willie B., he was living in a small concrete enclosure with bars in front. Willie B., who didn't have anything else to do, would often interact with the guests to their delight. However, it was clear to both guests and staff alike that Willie B.'s well-being was compromised by this environment. Not only was this habitat unable to provide what Willie B. needs, the entire environment did not suffice – he hadn't a social environment that a gorilla requires, he had very little cognitive stimulation, there was no perceptible ability for intrinsic reward, and so on. Willie B. suffered from a lack of any naturalistic functional equivalency in his habitat.

Rather heroically, the Atlanta Zoo recognised these deficiencies and a major overhaul of his environment (as well as the environments of other animals) was executed. The Atlanta Zoo became 'Zoo Atlanta', with a key initiative and mission to provide naturalistic habitats for the animals therein.

The idea behind providing naturalistic habitats is that, built into the design of a good naturalistic habitat, is a functional equivalency with what the animal would have in the wild. Animals can then exhibit more naturalistic behaviours, more naturalistic movements, and enjoy a sense of well-being and safety when interacting with habitat that they have adapted to throughout the course of their natural history. Adding to this is an improved guest experience whereby visitors are able to see an animal interact with their habitat in the manner in which they would in the wild. This allows for the zoo to educate the public on the ecological needs of the species they're exhibiting, while, at the same time, providing a better habitat and a safer habitat for the animals.

For Willie B., the zoo created a simulated African rainforest that attempted to reproduce the habitat that a lowland gorilla would find in Cameroon. Additionally, four other rainforest exhibits were created, allowing the zoo to bring in more gorillas. This allowed for social housing and social introductions with Willie B. The results were significant. After spending his life in a concrete enclosure, with no functional equivalency, Willie B. now had trees, foliage, and hills that he could climb. He had an expansive outdoor habitat where he could enjoy the rise and setting of the sun each day. The zoo took great pains to do regular behavioural assessments on Willie B. during this transition. They observed that he showed greater signs of positive welfare. He was more physically active. He explored. In his social interactions, Willie B. was breeding. In fact, he ended up fathering several offspring. At the same time, the public enjoyed watching Willie B. become so interactive with other gorillas (instead of the public). They learnt about the ecology of wild gorillas in a way that they could never experience from just reading a book.

They learnt about the effects of habitat loss on species like gorillas and the need for action to protect animal habitats. The transformation of Willie B.'s habitat is a great example of why providing a naturalistic habitat is an incredibly important step to providing good welfare to animals and managed populations.

The effects of functional equivalency on amphibious animals are another example of the benefits of a naturalistic habitat. Animals that naturally spend part of their time in the water and part of their time on the land in order to find food or locomote often find a lack of functional equivalency in an artificial habitat. Once again, novel habitat design offers a solution to this issue. Designs where the habitat is split between an aquatic habitat and a terrestrial habitat, with food possibly located in both places (creating a necessity to move across both habitats), can create a naturalistic functional equivalency for these animals. This can be seen in the habitat designs of penguins, seals, and other amphibious animals.

As zoos around the world have continued to move towards providing naturalistic habitats, the public has been able to watch this transformation while seeing very obvious signs of improved welfare for these animals. Parrots flying through a rainforest-type aviary provide both a better guest experience and better animal welfare than a parrot in a cage. Zebras and giraffes that are able to move through a savanna-type setting provide a better guest experience and better welfare than zebras and giraffes living in small pens.

Preference for naturalistic habitats is not relegated to mammals and birds. In fact, significant benefits to reptile well-being have been shown in naturalistic habitats (physiologically, behaviourally, and psychologically).[2] Studies with captive box turtles show a significant preference for naturalistic habitats, apparently driven by an innate sense of preference rather than prior experience.[3]

6.4   Degree of Physical Movement Possible within an Artificial Habitat

A man living in a small hut who wears a sports jacket has better welfare than a man living in a golden palace who wears a straitjacket. There is one simple reason for this *constraint*. Constraint, whether perceived or actual, is anathema to good welfare. When there is no freedom of movement, whether that movement be locomotive, or merely the movement of one's extremities in a way that is natural (dictated by how one would move in their natural environment), existence can be tortuous.

A key component of the anatomy of a natural environment is the degree of physical movement possible within that environment. As with every other component of an environment, animals have adapted to this over millions of years of evolution and natural history. These movements are not only necessary to survive one's natural environment, they are in fact, genetically inclined. Animals are driven by their genes to perform the movements they would normally perform within their natural habitat. Whether an animal is in their natural environment, or in an artificial environment, this inclination does not change. Indeed, one of the key reasons for poor animal welfare in an artificial environment is constraint – or a diminished degree of physical movement that an animal would enjoy in their natural environment. The ability to practise natural physical movements is foundational to self-determination and freedom of choice – one of our tenets of good animal welfare.

Allowing for Natural Movement

The ability to move one's extremities in the manner they would in their natural environment is a key component of an individual's welfare. These movements are directly

---

[2] See Warwick & Steedman, 2023.
[3] See Tetzlaff et al., 2016.

related to the specific environments where a species has evolved. An orangutan has long arms, and an operation to use those arms, because they move about through the treetops in Indonesian rainforest. A migratory bird has wings, and an operation to use those wings, which allows for long flights across a vast expanse of area. A dolphin has flukes and an operation to move those flukes, in a manner consistent with an environment where they can traverse large areas very quickly.

In many cases, an artificial environment presents significant challenges in allowing an animal to move in a way that they're naturally inclined to move. Logistically, containment itself often presents these challenges. A bird whose wings are pinioned to prevent escape is unable to move in a way that they're naturally inclined to move. A whale in an aquarium setting may be unable to utilise their flukes in a manner consistent with how they would move them in a natural setting. In such cases (with the knowledge of how the inability for natural movement may compromise welfare), it may be necessary to discuss whether this animal should be housed in this environment at all. If the animal is to remain in the artificial environment, methods to compensate for this lack of natural movement should be made. This, however, can present other significant challenges. Oftentimes, care technicians must look for novel methods of accomplishing this. This may entail temporarily moving animals to different locations where natural movement is possible. This may include an operating conditioning method where animals can perform their natural movements under given conditions. Whatever the method, animals should be consistently assessed to determine if these compensation methods are helping to improve the individual's welfare.

Allowing for Freedom of Movement

Animals, especially those that are locomotives, must make choices consistently throughout their day. Where to sleep, where to eat, which conspecific to approach, which conspecific to avoid, where to flee from a potential predator, where to hunt potential prey, where to find water, and almost every other aspect of an animal's individual survival, rests in their ability to make choices. The crux of being able to make such choices for a locomotive animal is the ability to have freedom of movement, or at least the same degree of movement possible that one has in their natural environment. In an artificial environment, there can be challenges to providing the same degree of freedom of movement that one has in their natural habitat.

There are two major components in providing this freedom: spatial and management.

- *Spatial component*

  Spatially, animals must be provided the amount of space that makes freedom of movement, or the degree of freedom of movement that they had in their natural habitat, possible. However, this can be impossible to provide in an artificial habitat for some species. For example, a chimpanzee that typically traverses 20 or 30 kilometres per day in the wild is not going to have that in an artificial environment. A dolphin that has the freedom of movement throughout the ocean is not going to have that in an aquarium setting. A bird that has the freedom of movement of the sky is not going to have that in any artificial environment. The design of artificial environments must take this into account and attempt to synthesise, as best it can, the effects of the space for that species. A transitory chimpanzee may need to move from one enclosure to another on a regular basis. They may also need the interior of their enclosure changed on a regular basis. Similarly, dolphin environments may need to be frequently changed to synthesise changing space.

Operant conditioning programs can also help. Some operant conditioning programs can promote movement and exercise – exercising the muscles that animals in artificial environments rarely utilise. This can minimise the effects of a lack of similar space given in a natural habitat.

- *Management component*

When managing an animal population, care technicians frequently need animals to move from one part of their habitat to another. However, an animal whose every movement is micromanaged by a care technician is going to have relatively poor welfare. Great effort must be made to provide as much self-determination and freedom of movement as possible. Again, some of this can be accomplished through operant conditioning programs, such as station training. Animals can then voluntarily participate in their management and have that degree of self-determination in the decision to participate. Husbandry schedules should also be made loose enough that animal movements do not need to be constrained to the point of micromanaging. Most of an animal's day should be given as 'free periods' – or times when animals don't have to be in one space versus another and are free to choose their own space.

6.5   Degree of Cognitive Stimulation Possible in an Artificial Environment

The manner in which an animal strategises the environment around them demands a cognitive ability. The use of this cognitive ability is an individual's cognitive stimulation. Through cognitive stimulation, animals are able to map out their habitat, devise strategies to acquire resources, create alliances with conspecifics, determine methods of safety, form breeding strategies, and parent offspring. As with everything else, animals have evolved their degree of cognitive stimulation relative to what is typically utilised in order for them to survive their specific environment.

In an artificial environment, most, if not all, of an individual's needs are provided. Food is provided, safety is provided, social groups are managed, and habitats are often unchanging. This can lead to the inability to utilise cognitive strategies. Therefore, cognitive stimulation can be lacking (and significantly less than what would be present in their natural environment). Without cognitive stimulation, animals can become bored, anxious, or depressed. They may not exhibit many of their natural behaviours. Their health can even be compromised (especially when this condition leads to a lack of typical movement). It is therefore critical that artificial environments provide the homologous degree of cognitive stimulation to what would be present within an animal's natural environment. Compensation for cognitive stimulation can be done through enrichment (see Section 6.8), management strategies, operant conditioning programs, habitat design, or novel methods.

Creating a Stimulating Environment

It is important to take into account that, in an artificial environment, most if not all animal needs are taken care of. Therefore, when managing animals in an artificial environment. protocols should be in place to allow animals to strategise as much as possible when obtaining the resources that are provided to them. This is not to say that resources should be deprived in any way. However, resources can be presented thoughtfully in order to promote cognitive stimulation. There are three major categories to providing animals with resources in a thoughtful manner: feeding, habitat, and socialisation.

- *Feeding*

  Animals should be fed in a manner that creates a functional equivalency in employing the strategies they would normally have to perform in order to acquire those resources. At its most basic, this rule would imply that browsing animals get fed up high, while foraging animals get fed down low. However, this should be implemented in an even more thoughtful and creative way. For example, a giraffe that, in the wild, has to search through several branches to find preferred food, may benefit from having this replicated in an artificial environment. A primate that would normally forage through the grass to find food, may benefit from having part of their diet scattered throughout the grass of their enclosure. A spoonbill that, in the wild, would be sifting through the water to find a preferred food source, may benefit from having their food placed in the water. Anathema to all of this is an animal in an artificial environment that is being fed directly. Anathema to all of this is an animal being fed in an unchanging trough. Anathema to all of this is an animal being fed in any manner that fails to utilise their natural feeding strategies.

- *Habitat*

  Habitats should be designed in a way that replicates as much of the cognitive stimulation that a habitat would possess in the wild. Animals that typically explore need places to explore. Animals that are nomadic and travel from one habitat to the next in the wild need a habitat that changes frequently and regularly. Animals that have natural habitats where they need to dig need an artificial habitat where they can dig, and so on.

  This has been shown to be true of all animals, regardless of their taxonomic classifications, be they mammals, birds, reptiles, amphibians, insects, fish, and any other taxon. For example, in one study, zebrafish showed signs of improved welfare (and less signs of anxiety and aggression), when novel spaces were created within their habitat – allowing for exploration.[4]

---

[4] See Graham et al., 2018.

- *Socialisation*

     How an animal strategises how to interact with their conspecifics is a key way in which they utilise their cognitive functions. Thus, socialisation is a major aspect of cognitive stimulation. Individuals ally together. They compete with each other. They fight with each other. They play with each other. They parent their young. They are reared by their parents. All of this requires cognition. Though we will discuss the *richness* of a social environment below, it is important to note that socialisation affects cognitive stimulation.

     An animal that normally has a very large social environment, but is living in an artificial environment where they have very little social interactions, is going to miss out on a large degree of cognitive stimulation. Despite valid attempts, *socialisation is almost impossible to reproduce without con specifics*. Therefore, it is key to provide animals with a rich social environment (see Section 6.6) in order for them to have appropriate levels of cognitive stimulation.

6.6   The Artificial Social Environment

What is and is not a proper social environment is critical to not just whether or not an animal can thrive but also whether or not an animal can even survive. Solitary animals naturally may not be able to survive living among conspecifics. Conversely, social animals may not be able to survive *without* living among conspecifics.

The degree in which an animal is inclined to live with or without conspecifics is the *richness* of its social environment. As such, we can define a **rich social environment** as an environment that most closely reflects the social makeup of a natural population of a species. Obviously, in managed care, there are considerable constraints to being able to house species in the same social environment as they would be in the wild. However, care technicians can attempt to replicate the richness of the social environment relative to the size of the habitat and relative to both the species and individual proclivities towards a social environment.

What is absolutely essential for care technicians to keep in mind is that a *created* social environment is an *artificial* social environment. In nature, social populations are created through births, lineage, philopatry, and other ecological means that might only exist in the natural world. However, in an artificial environment, social populations are created by human manipulation, so the stability one might see in the wild often doesn't exist in an artificial environment.

Beyond just the creation of a social group in the wild, nature has strategies for mitigating instability that are not present in an artificial environment. For example, in the wild, individuals can often leave their social group and go to other social groups. Social groups can also fracture and split into multiple groups. Animals living in artificial environments typically do not have access to these means. Therefore, care technicians may try to impose these means. For example, incompatible individuals may be removed from social groups and placed into other groups. Social groups may be subgrouped by care technicians and thus split into two different groups. Though sometimes necessary, these methods should not be approached lightly. In fact, performing these methods can have severe unintended consequences, especially for extremely complex groups. Without a deep knowledge of the inner workings of the social group, a care technician may inadvertently be removing individuals that are actually providing unseen stability or splitting alliances that might be allowing the group to function. Hence, keen observations of a social environment are essential for care technicians to provide the appropriate social richness to the animals in their care.

Creating a Rich Social Environment Relative to the Species

In the wild, an individual's social environment is determined by several factors. Among these are the availability of resources, the size of a habitat, and the amount of predators in the area. All of these factors are altered in an artificial environment. All food resources are provided for every individual. The presence of predators should not be a factor in a good artificial environment. Thus, neither should constrain population size. Available space, however, is a prime consideration in an artificial environment. For example, some species of primates may have social groups of 40 or 50 individuals. Obviously, there are very few artificial environments that could house such a population. Housing too many individuals with limited space can lead to considerable welfare concerns; including aggression (sometimes lethal), a lack of sanitation, and any other effects of general overcrowding. Beyond space, there are group dynamics that can determine the group size. Both space and population dynamics must be prime considerations when creating a rich social environment.

- *Constraints of space*

  The richness of a social environment is going to be constrained by space. An overcrowded social setting is not a rich social environment, no matter how social that species might be. Overcrowding is going to create issues of aggression and poor group dynamics that create profound limits on socialisation.

  Different species are going to have various constraints of space on population size. This is largely going to be determined by natural ecology. For example, howler monkeys, which, in the wild, are typically distributed throughout an extremely large area in an extremely dense forest, are going to have more constraints of space than animals that typically live in much closer proximity to each other. Once again, it becomes incumbent on the care technician to know everything that is relevant about the species and their natural ecology. Failure to do so creates a social environment that is either too sparse or too dense and can have critical welfare consequences to the individuals within.

- *Constraints of dynamics*

  The social dynamics of specific populations are going to create constraints on the richness of the social environment. These factors increase dramatically with species that have complex hierarchical structures. These species function most effectively when that hierarchy is stable, and when there are clearly defined social roles within that hierarchy. However, these same species can also be severely limited in the richness of their social environment when that hierarchy is unstable or when there are not clearly defined social roles. This instability can lead to aggression and can determine further limits on the amount of individuals within the population.

Creating a Rich Social Environment Relative to the Individual

Even though a species may be naturally social, or naturally solitary, each individual of that species is going to have their own social preferences and their own social proclivities to what is or isn't a rich social environment for them. These individual preferences might be based on their own social history, especially with animals that have been raised in human care. These also might just be individual preferences for how social they feel comfortable with. Combine this with the fact that some individuals have physical constraints, or psychological constraints, on how social their environment can be. This can also affect the ability for being introduced to other conspecifics.

All of these factors are going to determine how rich a social environment can be for *an individual.* Confounding this challenge when creating an artificial social environment is that each individual within a planned social group is going to have their own individual threshold for how social they can be. This becomes very challenging for care technicians, since they have to find common ground between the thresholds of each individual within the population they're trying to create. This may require creating multiple populations, temporary sub-populations, or it may require individual mitigation strategies to attempt to bridge the different challenges that may make certain individuals more or less social than their species typically is.

- *Social experience*

    An individual's social experience is going to have a profound impact on their social threshold. This is compounded when dealing with extremely social species that are also very territorial. Individuals that lack much social history may face major challenges attempting to function within a population. For example, some individuals may have been kept from conspecifics early in their life and therefore do not know how to behave in a social (or even territorial) setting. This can lead to these individuals being the victims of aggression, or being aggressors themselves, and can create critical instability within the population in an artificial environment. Even if there is no aggression, or other physical effects of these individuals living in a social setting, there may be undue stress and anxiety, which limits the individual's ability to thrive. More often than not, however, these individuals can be treated with mitigation strategies to slowly introduce them to a social setting and give them the experience they require to enter into a social group.

- *Physical challenges*

    Some individuals may have physical ailments that limit their social thresholds. These ailments may affect how they operate when being introduced to conspecifics. They may affect how they handle any aggression between conspecifics. These ailments may mean that they need to be under close veterinary care that won't allow the presence of other conspecifics. With all individuals, an individual's physical condition should be taken into account when determining an individual's social threshold.

* *Individual proclivities*

   In some cases, there may be no observable reason for an individual's social preference. Some individuals may just prefer to be more or less social than their species normally is. Often the factors that created these preferences may not be discernible to the care technician. However, these individual proclivities should be treated just as significantly as any other challenge to a rich social environment. Strategies such as utilising social enrichment, or breaking groups down into temporary smaller groups may encourage a less-than-social individual, to become more social. However, these proclivities may be lifelong. The anxiety or stress that socialisation can have on these individuals should always be taken into account.

6.7   Intrinsic Reward in an Artificial Environment

For us, as humans, it is difficult to imagine a world where there is no degree of intrinsic reward. The ability to walk outside when the weather is nice, just for the sake of the feeling of a cool breeze or sunshine, is an immeasurable part of our well-being. The ability to soothe ourselves with our favourite song is something we might feel that we cannot live without. The ability to learn a sport or learn an instrument may actually motivate our behaviours. For many of us reading a book for the pure pleasure of reading it, or watching a film that we've already seen, is a large part of where we find pleasure. These are all examples of performing actions by intrinsic motivation.

Animals are highly intrinsically motivated. Animals will perform actions for the sheer sake of performing them with no direct ecological benefit. Animals will explore an area that has no perceptible resources within it. Animals will play with conspecifics. Animals will perform very skilled (and energy-expensive) tasks when there is no actual need to do them. Animals perform intrinsically motivated actions because, by doing them, they increase their general well-being.

However, an artificial environment may lack the proper avenues for intrinsic reward. Oftentimes, this correlates with other factors of thriving that are compromised within an artificial environment. For example, a lack of the ability to move as one naturally does, can compromise movement for intrinsic reward. A dolphin that might swim at high speeds through the ocean, just for the intrinsic value of swimming fast, may not have this ability within an artificial aquarium environment where space is limited. A bird that may fly high just for the sheer sake of flying at that altitude may not have this ability in a canopy type of artificial environment.

Similarly, a lack of cognitive stimulation in an artificial environment can also compromise the ability for intrinsic reward. Animals in an artificial environment that is unchanging will not have the ability to explore as they might naturally do in the wild. Animals that perform problem-solving tasks in their natural environment might not have the ability for similar tasks in an artificial environment.

A lack of a rich social environment also can deeply compromise the ability for intrinsic reward. This can obviously especially compromise the ability to play. Social animals that either do not have other conspecifics within their artificial environment or do not have the *appropriate* conspecifics in their environment are not going to be able to play as they might naturally be inclined to do.

Knowing which avenues a species generally seeks for intrinsic reward is valuable knowledge for a care technician to possess. Some aspects of intrinsic reward cannot be replicated either due to logistics or due to safety. However, care technicians should look for ways in which these avenues can be *functionally* replicated. For example, problem-solving tasks can be functionally replicated with enrichment devices (see Section 6.8). Intrinsic reward involving movement may need to be accomplished in temporary environments. Habitats may need to be regularly changed to promote exploration. Whichever the method, all of a species' preferred avenues of intrinsic reward should have a functional equivalency within an artificial environment.

Creating Flow in an Artificial Environment

Oftentimes the ability of an artificial environment to match an individual's skills with the challenges inherent within the environment is lacking, resulting in an off-balanced environment. Either the individual within is unable to fully utilise their skills or there are challenges that the individual cannot mitigate. The results of this situation are animals that are either under undue stress and anxiety or bored. Finding that spot where the challenges of an environment meet the skills of the individuals within is an environment's flow.

Almost all other aspects of thriving in an artificial environment hinge on this balance – or flow. There are aspects of flow within the ability for an artificial habitat to provide the functional equivalency of a natural habitat. There are aspects of flow within an artificial environment's ability to provide for an individual's freedom of movement. There are aspects of flow within the amount of cognitive stimulation that is present within an artificial habitat. There are aspects of flow within each social interaction that a species is able to have with conspecifics within an artificial environment.

The crux of flow rests in the environment's ability to offer any intrinsic reward. Intrinsic reward is based on the ability to successfully utilise one's natural skills. For example, as a human, I might want to master playing a concertina or a mandolin because I have a brain that is inherently wired for music. The ability to play music stimulates this cognitive function. An instrument presents a challenge that I can successfully master with enough time and practice. However, I will never be able to climb to the top of a palm tree the way a chimpanzee can. I simple do not have the natural ability to do this. The challenge there exceeds my possible skills. Therefore, if I am forced to attempt it, I am bound to be frustrated or anxious. Thus, I find intrinsic reward in mastering a challenge that I have the skills to accomplish. This becomes my motivation. If I have nothing like this, I am left bored and my well-being is compromised.

In order to create flow in an artificial environment, the following factors are critical:

- *Knowledge of the species*

  Understanding the natural abilities, inclinations, and challenges that a species faces in the wild is key to understanding what defines the relative flow. Animals require a functional equivalency of these aspects of their ecology for flow to occur.

- *Knowledge of the individual*

  As with all aspects of animal care, there are species-level and individual-level considerations. Flow is no different. Individuals have unique preferences on what stimulates them and what skills they are most prone to utilising. An environment should be created that enables and promotes these individual preferences and the species-level needs.

- *Designing a habitat with flow in mind*

  Habitats must allow for the ability of a species to utilise their natural skills and abilities so that they can meet the challenges within their environment through their inherent faculties. Conversely, habitats must not present challenges that a species cannot mitigate.

- *Observing individual behaviour*

  Once in an artificial environment, individuals should be observed for signs of anxiety or boredom. When these signs are observed, care technicians should determine if there is an imbalance of flow within the environment and compensate appropriately.

- *Observing social compatibility*

  Social animals should be engaging socially with each other. If they are not, this may be a sign of social incompatibility. A lack of social engagement for naturally social animals means that animals are not strategising interactions and therefore not utilising the inherent social skills they possess. Signs of undue aggression, disinterest, or general fear can be signs of social incompatibility.

- *Maintaining a robust enrichment program*

  A robust enrichment program can be key to finding a balance of flow. Enrichment can provide challenges that can stimulate an individual's own preferences and species-level skills. We will discuss how to create a robust enrichment program.

6.8 Enrichment

When it comes to providing an environment where animals can thrive, among the most helpful tools in a care technician's arsenal is a robust enrichment program. A good enrichment program can provide the functional equivalency that is so often lacking in an artificial environment. Additionally, enrichment can be amended and tweaked to meet the needs of individuals, allowing an excellent way to provide individualised care when animals are facing unique challenges.

At its core, an **enrichment program** simply means providing prescribed stimulation for animals in artificial environments. Enrichment can range from very simple acts of stimulation, such as providing an animal's diet in a novel way, to more elaborate enrichment devices, such as building complex habitat structures, that may mirror an aspect of a natural habitat, so that an animal can engage with it in a stimulating and natural way.

Enrichment is both species-specific and individually specific. As such, enrichment is tailored to the general needs of the species but also can be constructed to meet unique needs of an individual. Enrichment may be manufactured to compensate for something that's lacking in an artificial environment, such as promoting movement in an environment that doesn't naturally lend itself to natural movement. Similarly, it can provide cognitive stimulation in an environment that is relatively routine.

The following are critical aspects of a robust enrichment program:

- *Enrichment is a state, not an event.*

    It is important to not view enrichment as a box to tick each day. It is, in fact, an environmental state. A good environment is an enriching environment. Enrichment does not start when an enrichment item is given and stop when the device is expended. Rather, enrichment items should be small parcels of an overall enriching environment. For example, an animal who is given 15 minutes of stimulation in a sterile and unstimulating environment, is still living 23 hours and 24 minutes of each day in an sterile and unstimulating environment. This animal is not thriving and lives in a state of negative well-being. Thus, the entire environment needs to be stimulating. This can be accomplished with an enrichment program that looks at the overall state of stimulation that an animal is receiving.

    A truly robust enrichment program includes a multi-pronged approach where an array of enrichment items are used in conjunction together to create an overall program. This ensures that an animal is living in an enriching environment and in an enriching *state*, rather than just having sporadic enriching events. This is the definition of a *robust* enrichment program.
- *Enrichment is planned.*

    For an enrichment program to be effective, it must be carefully planned and executed according to this plan. It should employ a strategy that stimulates key aspects of thriving and, in such a way, that the animal lives consistently in an enriching environment.
- *Enrichment is assessed.*

    A robust enrichment program includes a regular assessment in order to determine if that program is actually working to create a stimulating environment.

Enrichment Imperatives

In order for an enrichment program to effectively create a stimulating environment for each animal within, it must achieve specific imperatives relative to the needs of the species and the needs of the individuals. Fulfilling these imperatives can significantly aid in compensating for what is lacking in an artificial environment.

It is important to note that most enrichment devices or items will not fulfil all imperatives at once. However, it is important to determine which imperative is fulfilled with each item. This way an array of items can be utilised to create the overall enrichment program and thus create an enriching and stimulating environment.

A well-rounded, and truly robust, enrichment program creates an environment where all the residents within the environment enjoy a functional equivalency of the

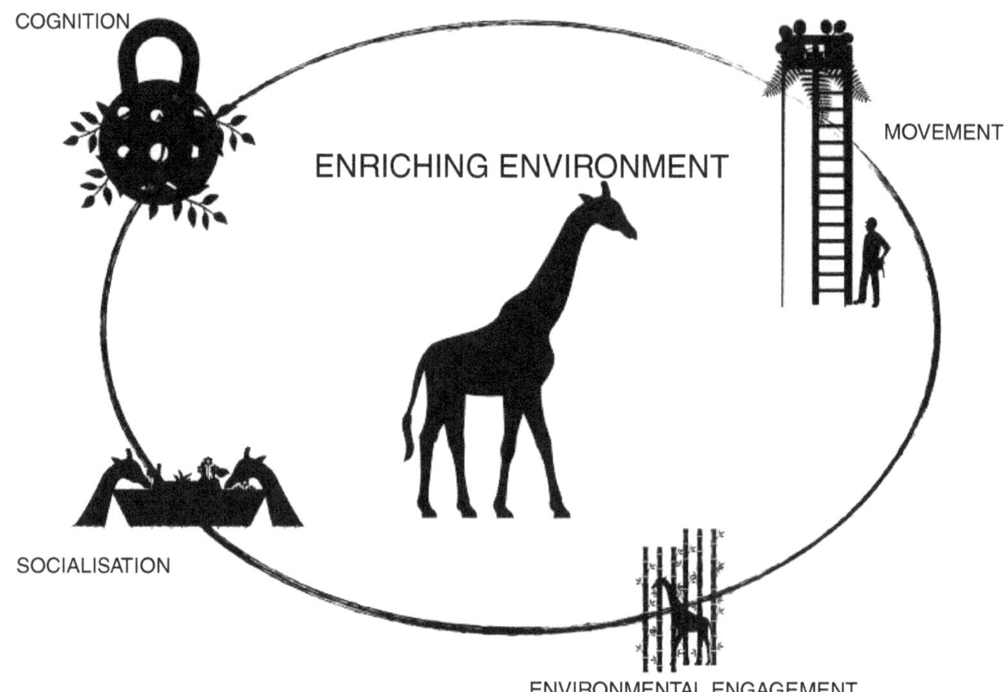

Figure 6.3 An enriching environment is a state whereby cognition, natural movement, environmental engagement, and socialisation are all stimulated.

stimulation they would receive in a natural habitat. Anything that produces the results of these imperatives, in a prescribed manner, constitutes enrichment (Figure 6.3).

- *Promoting cognitive stimulation*

When enrichment is aimed at cognitive stimulation, it should attempt to mirror the cognitive stimulation that the species would encounter in the wild. For example, in the wild, a chimpanzee will devise novel methods of obtaining resources such as making and using tools. Effective cognitive enrichment for chimpanzees incorporates placing a resource in a way that requires a chimpanzee to devise a novel way of obtaining it, such as requiring the use of a tool.

Often cognitive stimulation involves an enrichment where the outcome is unknown to the animal. This type of enrichment promotes exploration. The challenge to this type of enrichment is to promote the desire for an animal to engage with it. This can be accomplished with scents or just a novel device. Once again, this depends on both the species and the individual.

Finally, for an enrichment to truly provide cognitive stimulation, it needs to last for an extended period of time. A puzzle box that can be figured out in seconds is not cognitively stimulating. The longer an animal engages with enrichment, the more stimulating it is.

- *Promoting natural movement*

Depending on the species, some degree of natural movement may be impossible within a natural environment. Sometimes, even when natural movement is possible, it might be difficult and therefore not encouraged by the environment. It is in these situations where enrichment is an invaluable asset. Enrichment can provide a functional equivalency to an animal's natural movement. Properly executed enrichment allows an animal to exercise natural motions. This ultimately can benefit psychological, behavioural, and physical well-being.

Sometimes it is a simple act of management. For example, just by presenting a dietary item in a manner that encourages an animal to move as they would when feeding in a natural habitat, natural movement is achieved. For example, a giraffe feeding platform may be constructed that requires a giraffe to stretch its neck in order to reach the food.

Sometimes the solution may be more complex, including the construction of novel habitat structures, or even operant conditioning programs where certain movements are encouraged and rewarded.

- *Promoting environment engagement*

An animal may have a vast habitat and/or a beautifully constructed environment but not benefitting from it. If an animal is not engaging with the environment, then. No matter how well put together, the environment is in vain. A lack of engagement with the environment can be due to both external factors, such as an improperly constructed environment, and internal factors such as a physical condition, behavioural condition, or psychological condition. In all cases, enrichment can help mitigate these challenges and promote active engagement with the environment.

Environments that are frequently changed promote exploration. Environments where food can be placed in novel ways as a reward promotes active engagement. Interesting sounds coming from the environment can promote engagement. Interesting scents can promote engagement. Even housing animals in a mixed-species environment (when appropriate) can encourage engagement with the environment. All in all, determining an enrichment that promotes environmental engagement is oftentimes very specific to a situation and must be determined by the care technician with regard to the species and the individuals they're taking care of.

- *Promoting socialisation*

There are enrichments that promote socialisation by requiring conspecifics to work together to achieve a certain goal. For example, there are large spin feeders that can be used with primates, where if one individual is spinning the device to obtain the reward, the other individual either must spin it in conjunction with the conspecific so that they can both achieve the reward or must wait until the conspecifics actions align with their ability to also obtain the reward. Either way, socialisation is promoted and achieved. This type of enrichment encourages socialisation in a positive way. Group cohesion, as well as building bonds and companionship between two individuals within a group, are all promoted. This can be utilised when creating new social groups or adding individuals onto existing social groups. It can be used to stabilise social groups that may have become unstable.

Social enrichment strategies obviously need to be constructed relative to the social nature of the species. In all cases, social enrichment should be constructed to achieve the species-specific richness of the social environment desired.

Assessing Enrichment

As stated earlier, assessing enrichment is essential to creating an enriching environment. Despite even the best planning, some enrichments are not effective in achieving a desired goal. A proper assessment tool can allow care technicians to see where the enrichment is falling short.

Additionally, by assessing enrichment, care technicians can understand which of the above-mentioned imperatives the enrichment is strongest at promoting. This can help shape an enrichment schedule so that all of the imperatives are met regularly.

The following is an example of a tool that can accomplish an effective assessment of enrichment.

Cognitive Stimulation

1. Are individuals engaging with enrichment for longer than 15 minutes?
2. Are animals achieving a goal with an unknown outcome – meaning is the reward unpredictable. Or something that the individual is expecting?

Score '3' for two 'yes' answers, score '2' for one 'yes' answer, and score '1' for zero 'yes' answers.

Score _________

Promotion of Movement

1. Are animals required to be locomotive to achieve enrichment goals – meaning is the enrichment causing them to have to traverse a portion of their environment?
2. Are animals required to move in a novel way to achieve enrichment goals – meaning, is the presence of the enrichment forcing the individual to move in a way they normally do not?

Score '3' for two 'yes' answers, score '2' for one 'yes' answer, and score '1' for zero 'yes' answers.

Score _________

Environmental Engagement

1. Are animals interacting with a standing element in their environment in a novel way – meaning, is the presence of this enrichment causing the individual to act differently towards an object that is normally in their environment?
2. Are animals utilising two or more elements in their environment to achieve a goal – meaning are they having to use one part of their environment as a tool?

Score '3' for two 'yes' answers, score '2' for one 'yes' answer, and score '1' for zero 'yes' answers.

Score _________

Social Engagement

1. Are two or more conspecifics engaging in the enrichment at the same time – meaning, are they sharing a device or structure?[5]
2. Are conspecifics working together to achieve a goal (as opposed to just being in proximity together) – meaning, does one conspecific's action affect the manner in which another obtains the reward?

Score '3' for two 'yes' answers, score '2' for one 'yes' answer, and score '1' for zero 'yes' answers.

Score __________

Total Score __________

>9 – Highly effective enrichment
7–8 – Effective, but may require some adjustments
<7 – Ineffective enrichment, requires major adjustments or cancellation.
Which category, or categories, is/are strongest according to the scores?________

Implications

The key to promoting positive welfare in an artificial environment rests in the ability to create a *functional equivalency* with what an animal would be faced with in a natural environment. A functional equivalency does not mean replicating all aspects of a natural environment. In fact, it doesn't even mean attempting to mirror a natural environment at all. Rather, a functional equivalency means that the variables of the artificial environment produce the same effects of the variables of a natural environment in terms of movement, socialisation, cognitive stimulation, and intrinsic reward.

Enrichment is one of the most powerful tools in a care technician's arsenal that can provide functional equivalency to an artificial environment. Enrichment is a scientific process that includes careful planning, a knowledge of the species, a knowledge of the individuals within the population, and a method of assessment.

Case Study: A Long and Winding Career Leading to Primate Canopy Trails at the Saint Louis Zoo

Heidi Hellmuth, Curator of Primates and the Director of the Center for the Conservation of Congo Apes, Saint Louis Zoo

On the Long and Winding Career
I am an anomaly, or at least unusual in the animal care field, in that I didn't 'always' want to work with animals growing up, and I never volunteered or interned working

------

[5] Assessment is relative to the natural social proclivities of the species.

Figure 6.4 Heidi Hellmuth. (Photo Credit: author).

with animals before getting my first full-time position. Even though my path wasn't traditional, I did learn some great life lessons along the way that helped me to achieve many of my career goals.

Growing up, I remember thinking I would 'know' what I wanted to do for my career. So when I was in high school and still hadn't had that revelation or aha moment, I was a bit nervous. I did some student teaching in a hearing-impaired elementary school class during my junior and senior years of high school, and enjoyed that, so decided to go for an education degree in college. Even though I thought I would enjoy teaching, it didn't feel like I had found my real niche.

Fast forward to the summer after my freshman year of college. I lived in south Florida at the time, and my parents and I were visiting the Florida Keys, including a stop at a small dolphinarium. After watching a dolphin show, I was able to spend some time talking to one of the trainers, and pretty much that interaction changed the course of my life. I decided that day that being a marine animal trainer was what I wanted to do, but I was at least somewhat realistic enough to know that I probably didn't have enough information to make such a big decision based on that one event. So, I spent that summer and the beginning of the next school year learning whatever I could about being a marine animal trainer including finding out about and joining the International Marine Animal Trainers Association and contacting people in the field to try to set up a summer job for the next year.

While I thought I had summer employment set up for after my sophomore year, it ended up falling through at the last minute which was pretty devastating. I was literally on my trip back to south FL from college in OH when I found out, when I had stopped at my brother's place in northern FL for a visit. So I wallowed for a bit, then got on the phone (no PC's and internet back then) and called the aquariums in the vicinity to see if any were hiring for the summer. Two were Sea World in Orlando and Marineland of Florida in St. Augustine. I managed to get interviews at both and was offered a position as 'apprentice dolphin trainer' at Marineland of Florida, doing all the grunt work, and making a whopping $3.25 an hour. But it was invaluable experience, and it solidified my interest and desire to pursue this as my career.

On the Animal Care Field

The animal care field is incredibly competitive and can be tough to break into. Once I graduated from college, I targeted the northeast US, since I had relatives in the area I could stay with for free while visiting/interviewing and there were some jobs advertised. My goal was to move wherever I had to for that first job, to get my foot in the door of the industry and gain experience. I had two interviews, at Mystic Marinelife Aquarium in CT and New England Aquarium in MA, but I also contacted several other aquariums and oceanariums in the region and asked if I could meet to talk about what they look for in a trainer as I was trying to break into the field. Every one of them said yes, and those conversations were invaluable in gaining a better understanding of what employers wanted and making contacts that might be valuable in the future. This was a great lesson for me that taking initiative and putting yourself out there, professionally and respectfully, may be daunting but can be incredibly beneficial. And I was also fortunate to be hired by Mystic Marinelife Aquarium which started my full-time career in the animal care field over 35 years ago. I've never looked back.

While I've not looked back, my career path has also not been linear. After about seven years working with marine mammals, I shifted my focus. What would later be called behavioural husbandry or behaviour management was starting to emerge, especially the training of more species of zoo animals for management and husbandry behaviours. And in some cases, marine animal trainers were pioneers of expanding the training of other zoo animals and the keepers who cared for them. I thought this sounded like about the best job in the world, so my new goal became plotting a course to get there. I knew that while I had gained a lot of great experience so far, I was not ready to get or excel at a job like this – yet. This is when I first started to realise that careers can take many twists and turns and began to think about gaining skill sets towards a future goal. In this case, I saw some people moving into zoo animal training hitting obstacles, such as not having experience with non-marine animal taxa and not understanding the realities of the animal keeper job. So, I set out to gain skill sets in both areas to help prepare myself for my new career goal.

On First Steps

My first step was to get a job as an animal keeper, to help me learn new taxa and the reality of the keeper role. It wasn't easy to get a keeper job, my years as a trainer actually worked against me as some employers thought I might not be happy in a keeper role. But with some persistence – and taking a one-third pay cut from my trainer position – I was hired as a carnivore and hoofstock keeper. It's hard to describe how beneficial this experience was in my growth. I thought I understood the differences between a trainer and keeper job, but I did not. I learnt about so many kinds of animals, from nervous and flighty to large and dangerous, and started to see how increased behaviour management could – and couldn't – fit in to a keeper routine. And, oh yeah, I worked really hard but had so much fun. This experience was a huge piece to my career puzzle and was invaluable in later roles.

For the next dozen years, I continued to gain skill sets that would help me in my hoped-for behaviour management position in the future. In addition to zoo keeper experience, I was able to help develop and manage a wildlife education show and gained my first zoo supervisory and management experience. I had a chance to work with a variety of taxa, both in zoo and show situations, and got my first experience with non-mammalian marine species including fish, rays, and cephalopods, which was pretty cool. The experience, skills, and networking paid off when I was hired in 2006 as the first ever Curator of Enrichment and Training at the Smithsonian's National Zoo, where I helped to implement comprehensive enrichment and training programs at the National Zoo and Conservation Biology Institute and served as a training and enrichment mentor and resource for the animal care staff at both facilities. My six and a half years there was truly one of the most special times in my career. I learnt so much, was able to experience an incredibly diverse array of animals, and hopefully helped the staff grow and develop more tools to give these animals the best care and welfare.

On the Main Lessons Learnt

Firstly, taking initiative, like when I learnt about and joined the International Marine Animal Trainers Association even before I was in the field, and reaching out to facilities to ask if they'd meet about what they look for in staff. For the latter, I only did so once, and if I didn't get a response I let it be, but I was very pleasantly surprised that the vast majority of people I contacted were happy to talk to me. Also not getting stuck in a linear career path if your interests change along the way. In this case look at what you need to help you get to your new goal, and don't be afraid to take jobs that are to gain skill sets that may not be your exact career goal. When I looked for these skill set opportunities, my rule of thumb was to plan to commit 3–5 years to that role to learn it well but also to give my employer a return on the investment they made in training me. And lastly, even though I don't go into it in detail here, is getting up off the mat when your career kicks you in the teeth. It's happened to me more than once, and I could have let it keep me down, but I didn't. Get back up, take a good look in the mirror on your role in whatever happened to improve yourself, then start working to get that next step in your career.

On the Primate Canopy Trails

In my current role as Curator of Primates at the Saint Louis Zoo, I've been able to take the knowledge and skills I've learnt throughout my career to help support our amazing team of animal keepers and zoological managers in the primate unit. One

Figure 6.5 The Majestic Forest within the Primate Canopy Trails. (Photo credit: author).

Figure 6.6 Heidi Hellmuth observing monkeys in the Primate Canopy Trails. (Photo credit: author).

of the accomplishments in my career I'm incredibly proud of is an amazing exhibit called Primate Canopy Trails (PCT), which transformed the lives of the monkeys and lemurs at the Saint Louis Zoo.

The PCT was opened in 2021 after about three years of concept, design, planning, construction, and animal acclimation. It was very much a team effort involving zoo staff (primate team, architecture and planning, facilities maintenance, graphics and interpretation, education, horticulture, and more), the architecture firm, general contractor, and many subcontractors. I've been involved in new exhibits at other times in my career, and I can truly say that the PCT team was the most collaborative that I've ever experienced.

Some history of why PCT was needed and built. The Zoo's Primate House is over 100 years old, and definitely has some limitations that come with being such an old building. Although there have been numerous renovations and improvements made over the years, the Primate House still had constraints on size of habitats as well as the fact that it wasn't originally built with any outdoor areas for the animals. In the 1990s, some outdoor wood and mesh structures were built along one side of the building, which did allow a limited number of animals to rotate outdoors – which was great – but their construction didn't allow all of the species to use them and they were limited in size and number.

The PCT's main, driving goal was to allow all the monkeys and lemurs to go outside, some for the very first time. Everything about the planning was designed towards that main purpose. Secondly, we wanted to save and include some beautiful, large trees on the site in some of the primate habitats. And finally, we wanted PCT to be a rotational exhibit – meaning animals could regularly move around to different locations – to maximise the opportunities for the animals to explore and enjoy as many spaces as possible. The rotational goal would be achieved by designing a network of tunnels that connected the new outdoor habitats to the Primate House building, to each other, and to a new Primate Care Center (PCC) that was built to offer additional indoor shelter options for the animals in the PCT habitats.

The historic Primate House has 12 indoor habitats (numbered 1–12 on the diagram), and the new PCT has eight outdoor habitats (numbered 13–20 on the diagram). Each habitat has direct tunnel access to an indoor area for shelter as needed. Four habitats connect to the existing Primate House indoor areas, and four habitats have shelter via our new PCC building (numbered 21–24 on the diagram). The PCT includes 12 tunnels, including the 140-foot-long 'Home Run' tunnel, which is the longest in the project, and which provides a direct connection between the Primate House and PCT.

Outdoor access, natural trees, rotation, and tunnel systems were our foundation, so with these guiding principles in mind the group started fleshing out the concept. We needed to make sure that PCT was designed with the needs of the animals, keepers, and guests in mind.

For our guests, we wanted them to be able to see the primates in new and unique ways, to help build connections that might inspire conservation actions. The main entrance to PCT has a large climbing area for the human primates to experience, using many of the behaviours and abilities like those of their monkey and lemur counterparts. There is a climbing tower, rope crawl tunnels, and an acrylic tube that goes directly through one of the primate habitats at an elevated level so guests may experience monkeys or lemurs all around them. For the less adventurous, there is a glass tunnel at ground level for all guests to get this eye-to-eye immersive experience. Another significant component of PCT is an elevated boardwalk that rises to about 12 feet in height and curves around to offer treetop level viewing of four of the PCT habitats. The boardwalk also helps protect the trees in the habitats by keeping people from walking at ground level on their root systems. Throughout PCT we talk about the challenges facing wild primates and what people can do to help, with the

Figure 6.7 The Primate Panorama within the Primate Canopy Trails. (Photo credit: author).

main message being that healthy, non-fragmented forests are important to all primates, including us. We also invite guests to learn how we care for the primates, with viewing windows into the PCC to see the keeper workspace and a glimpse into one of the animal bedrooms. We have interactive graphics that talk about the importance of animal enrichment and two browse gardens for the animals that are visible to guests that we interpret as well.

Our keeper team was an integral part of the planning and building of PCT. Not just tapping their expertise with the animals in helping design the habitats, but in making these areas work well for the keepers as they did their daily tasks. An example – the original design had locking hose bibs inside each animal habitat, where they keeper would bring in a hose, unlock the hose bib, hook up the hose, and clean as needed. This was, to me at least, a bit nuts. So with feedback from our team, we requested that each habitat vestibule – which was included for secondary containment safety – had a hose and spigot and a waterproof electrical outlet. This allowed the hoses to stay hooked up and ready to go each day versus having to be carried around and hooked/ unhooked at each habitat. Since keepers may share space with some of our smaller species, we also had custom designed 'hose port doors' built into the vestibule to habitat door that could be opened to put the hose through so the door could be closed as keepers worked, but which could also be locked closed when not in use for animal containment and safety. This is just one example, but we had the keepers help assure that they could safely and relatively easily access the habitats and buildings for their daily cleaning, even down to helping to design the most efficient floor drain system for the PCC. Without their involvement throughout the process, their jobs might have been inadvertently made more difficult.

But of course the majority of our focus was on the animals' comfort and well-being in the new PCT. Again the multidisciplinary team working on PCT helped look at this from many perspectives, which made our efforts more well-rounded and comprehensive. The new, large PCT outdoor habitats are all different shapes and sizes, but they all have key features for animal welfare. Starting with two large, elevated shelter boxes. These provide three-sided coverage from sun, wind, or other elements. They each have a divider so they can be one large space or two smaller ones. And on the back wall there are heat mats to give the animals a warm spot on a cool day. The design of these ended up working better than we imagined, as animals not only use the shelter, but the ladder structure that goes from the ground to the shelter, the roof on top of the structure which seems to be a favourite resting area, and there's even an individual perching area on the top of the post that holds the shelter up – they've really been a great success! Each habitat also has deadfall holders, basically plastic underground sleeves that hold trees/branches for locomotion and perching. These are held in place with about 1 inch of concrete at ground level, and when the deadfall needs to be replaced, the concrete is chipped out, new wood installed, and a new concrete cap poured. And as previously mentioned, three of the habitats also have large, natural trees that the animals can use and enjoy. Every habitat has one or more mister fans, and custom designed shade for each as well. Two lixits for fresh water, one just above ground level and one elevated, provide drinking options at multiple levels. The tunnel systems are not simply highways for animals to use to get from place to place, they are also destinations on their own and the animals regularly have access to these in addition to habitats and holding areas. And every habitat had stainless steel D rings welded to the structural steel as part of the design, offering numerous attachment points for exhibit furniture and perches, as well as feeders and enrichment devices.

While we didn't get everything right, and have learnt from the things we missed, PCT is one of the most versatile and (humbly) amazing exhibits I've seen, from the perspective of its effect on the lives and welfare of the animals. One example to show this impact is our large troop of Guereza colobus monkeys. This group included animals from about one year to the late 20s, and not one of these animals had ever been outside before PCT. Being our largest social group (which has ranged in size, but at this time there were eight animals), they had always been in the largest Primate House habitat which was 850 square feet/12,325 cubic feet (plus holding areas). That had been their home for over a decade. In about the first year of operation of PCT, due to the tunnels and rotational design that included the indoor Primate House and PCT, the colobus monkey group had experienced eight total habitats (two inside and six outside) which had a combined square footage of 6,270 square feet/128,982 cubic feet. And they each got to experience the sun, rain, wind, fresh air, and sights and sounds of being outside for the first time in their lives.

I feel very fortunate to have been involved in something that has had, and will continue to have, such a positive impact on so many animals. And the collaboration of our multiple stakeholders made PCT better than it would have been from any of us individually. Long after I retire, I'll be happy knowing that animals, keepers, and guests are still benefiting from this unique and amazing exhibit.

References

Diaz, J. and Reese, A. T., 2021. Possibilities and limits for using the gut microbiome to improve captive animal health. *Animal Microbiome*, *3*(1), pp. 1–14.

Graham, C., von Keyserlingk, M. A. and Franks, B., 2018. Free-choice exploration increases affiliative behaviour in zebrafish. *Applied Animal Behaviour Science*, *203*, pp. 103–110.

Warwick, C. and Steedman, C., 2023. Naturalistic versus unnaturalistic environments. In Warwick, C., Arena, P. C. and Burghart, G. M., *Health and welfare of captive reptiles* (pp. 487–507). Cham: Springer International Publishing.

Part III

Internal Factors

Curatio Fundamentorum **5:** Physiological and psychological mechanisms allow an individual to survive. The health of these internal factors determines an individual's ability to thrive.

Psychological and Emotional Health

Figure 7.1 Python in the herpatarium at the Chattanooga Zoo. (Photo credit: author).

7.1 Stormy Seas

A mariner faces a fierce and unforgiving sea. Other than his compass, he has no visual bearings signifying his direction. There is no line on the horizon. Waves the size of foothills lift and drop his ship into the bowels of the glacial waters. Horizontal rain mixed with chunks of ice pelt his face.

The Mariner clings to the mast as he struggles to read his compass. He is happy to see that he is still headed north. The rain and ice become more intense. The waves become taller. The ship rocks wildly, dipping a side of the deck into the ocean. The swaying of the ship becomes ever more pronounced.

The sails blow about like an open book in the wind. As The Mariner continues to cling to the mast, the boat tips to the point where his face comes within inches of the edge of a wave. All at once, a memory from long ago fills The Mariner's thoughts.

He remembers being young, on a boat with his father. He remembers a large wave coming towards them. He remembers the wave lifting the small boat. He remembers being thrown from the boat and falling into the water. He remembers the fear of being in the water. He remembers another large wave coming towards him. He remembers the wave picking him up and dragging him down deep into the water, twisting and turning him in the murky depths. He remembers the feeling of helplessly trying to paddle himself out of the fierce undertow. He remembers the confusion of not knowing the direction of the water's surface and not knowing if he was upside down, right side up, or somewhere in between. He remembers believing that he was going to die.

The memory fills him with an immense panic. The panic confuses him and detaches him from his present situation. His heart pounds. He gasps for air. He stops thinking straight. The panic overtakes any knowledge he once possessed of his ship. The panic overtakes his instincts on how to right his ship.

Another wave lifts the ship, temporarily righting the mast. Wasting the opportunity, he scurries around furiously and fruitlessly not knowing what to do next. He forgets to let out the sail. He forgets to hold the wheel. He stares motionless at the chaos in front of him. His indecision is costly. A wave sweeps over the bow, crashing over the sail, almost tearing its fabric and ruining all chances for the man's survival.

At this point, the instinct to survive takes over. The Mariner attempts to regain his composure. Another memory begins to enter into his consciousness. A memory appears of that fateful day when he fell off the boat. He remembers his father reaching out and grabbing him out of the water and pulling him back onto the boat. He remembers his father telling him that the best way to handle turbulence in the water is to swim with the tide, not against the tide.

The Mariner straightens his head and looks out at the stormy horizon. He stares at the rough sea, furrowing his brow. He now knows what to do. He releases the mast, lets out the sail, and allows the wind to take the ship where it's determined to take it. Almost immediately, the ship stops rocking. The rocking is replaced by a rapid forward motion. The vessel rides the stormy waves to the edge of the tempest.

Soon The Mariner and his ship are past the raging waves and the stormy skies. As The Mariner enters a calm space in the sea, he sits down on the deck and catches his breath.

The Persistence of Memory

In our nautical tale, The Mariner is armed with an awareness of his present situation, a knowledge of how his ship functions, and an instinct to survive. However, these tools can get overtaken by the shackles of a negative emotional state and negative reactions, or in this case, inaction, to the given situation.

The Mariner's environment has become extremely turbulent (quite literally). His perception of the situation is coloured by memories of previous situations. At first,

The Mariner is recalling a very negative experience. He recalls a feeling of panic. He recalls a feeling of impending doom. This memory creates a panicked emotional state, which causes him to not react appropriately to the stormy sea. It is not until he also recalls a knowledge that was passed to him from his father on how to deal with this situation in the future. This memory allows him to calm his emotional state, allowing him to face the given situation and survive it.

The small moment in The Mariner's life underscores how a neurological process occurs – how it creates our perception of the external variables we face, and how it leads to actions (either positively or negatively) affecting our survival and our well-being.

The Interplay between Internal Variables

The internal variables of an individual should always be looked upon holistically. This means that there is constant interplay between the physiological, the behavioural, the psychological, and the emotional. Each will influence the other at various degrees of significance, depending on the situation and the variable itself. However, for the purposes of this book, we will examine three categories of internal variables separately. The neurological (meaning psychological and emotional), the physiological, and the behavioural. However, because this interplay is so significant, we will need to consistently spill over into the other categories of internal variables.

However, it is impossible to separate out the two neurological internal variables – the psychological and the emotional. One is *always* defined by the other. Therefore, we will define both, but look at them as one larger neurological process.

7.2   Defining Psychological and Emotional Health

In order to truly understand the interplay between the psychological and the emotional, we must first define exactly what each one is. Psychological health and emotional health are two foundational aspects of overall well-being in animals. While they are closely related, there are distinct differences between the two concepts. As we have already explored when discussing psychological thriving and emotional thriving, the psychological relates to how we mentally perceive the world around us, while the emotional involves how we react to these perceptions – both of which determine how we interact with the world around us, or our behaviour.

By its definition, psychological health refers to the state of an animal's mental well-being, encompassing various cognitive and behavioural aspects. It involves the animal's ability to cope with stress, adapt to its environment, and maintain a positive sense of well-being. Psychological health is influenced by factors such as social interactions, environmental enrichment, and the fulfilment of basic needs. Animals with good psychological health exhibit behaviours indicative of contentment, curiosity, and a balanced emotional state.[1]

[1] See McMillan, 2005.

On the other hand, emotional health is really related to the animal's ability to experience and express emotions in a healthy and adaptive manner. Emotional health encompasses a range of emotions, including joy, fear, sadness, and anger. Animals with good emotional health are capable of experiencing positive emotions, managing negative emotions, and displaying appropriate emotional responses to various stimuli.[2]

While psychological health and emotional health are interconnected, they differ in terms of their scope and measurement. Psychological health encompasses a broader range of cognitive and behavioural aspects, including cognitive abilities, problem-solving skills, and social interactions. Emotional health, on the other hand, is more focused on the animal's emotional experiences and expressions.

As we will see, measuring psychological health in animals can be challenging due to the subjective nature of mental states. Researchers often rely on behavioural observations, physiological measurements, and cognitive assessments to evaluate psychological well-being. For example, in the context of animal captivity, researchers may assess an animal's psychological health by observing its engagement in natural behaviours, its ability to cope with stressors, and its overall level of arousal.

Similarly, measuring emotional health in animals requires careful consideration of their specific emotional experiences and expressions. Researchers may use behavioural indicators, such as facial expressions, vocalisations, and body language, to assess an animal's emotional state. Additionally, physiological measures, such as heart rate variability and cortisol levels, can provide insights into an animal's emotional well-being.

The psychological and the emotional work together in a mental process. There are two categories of the mental process. Firstly, there is the a priori, or everything that we psychologically understand *before* our perception. The a priori has to do with what we know instinctively. For example, the sound of a predatory animal usually creates an instinctual perception in their prey. The perception of danger is instinctively understood. This perception creates an emotional state.

Secondly, there is the a posteriori, or everything we psychologically understand *after* our perceptions (or what we know from experience). A posteriori perceptions have to do with memory and recall from previous events, or even what we know from what we've learnt. A posteriori psychological perceptions are highly individualised and depend on that individual's unique and specific history, and unique and specific experiences. For example, an individual history full of traumatic interactions with a specific perception is going to elicit a very negative psychological perception and thus a negative emotional state.

The recognition of the emotional states of animals have unfortunately gone largely disregarded for the bulk of the history of animal science. However, neuroscience has undoubtedly shown again and again that animals undergo **affective neurological processes** that create emotional states.[3] Obvious and irrefutable examples of affective

---

brain processes in animals include fear, panic, anger, and play. However, as animal behaviour continues to be studied, we begin to see that far more complex emotional states exist in animals, such as grief, elation, seeking, and curiosity.

As with humans, an animal's individual emotional state is reflective of their general well-being at any given moment. As such, care technicians must assess the emotional state of the animals within their care to properly assess their general state of animal welfare.[4]

Action-Oriented Systems

The interplay between the *psychological* and the *emotional* exists in what is called an action-oriented response.[5] An **action-oriented system** is a process whereby a mental perception causes an action either by motivational urge or involuntary response (Figure 7.2).[6] Such a system works like this:

- An individual interacts with an external factor within an environment.
  - The external factor may present danger, resource, or may just be a neutral presence.
- The interaction between the external factor and the individual creates a perception within the individual's mental process.
  - This perception determines the individual's attitude towards the external factor – positive, negative, or neutral.
  - The perception may be based on an instinctual perception or past experiences through the individual's memory/recall process. For example, if an individual encounters the sound of a predator, that individual may have an instinctual perception, whereby the sound is perceived as negative and something to avoid. Or, that individual may have a memory/recall perception, also eliciting an urge to flee the area. Similarly, the scent of a preferred food may have an opposite perception – again, either by instinct, memory/recall, or both, and create a perception of positivity.
- Once that perception is realised, the neurological process of an action-oriented system is initiated, where the individual is given to a particular action based on the perception.
  - If the perception is negative, perhaps evoking fear, the action-oriented system may dictate a **fight or flight response** (a physiological and behavioural response to danger). Involuntary actions such as elevated heart rate, rapid breathing, and quick reflexes, would be an action-oriented response to this perception, as would the motivational urge to flee the area.
- Coupled with the action-oriented response is an action-oriented emotional state.
  - For example, negative perception of danger could create an emotional state of fear or even panic.

[4] See Maple, 2015.
[5] See Mandik, 2005.
[6] See Mellor, 2012.

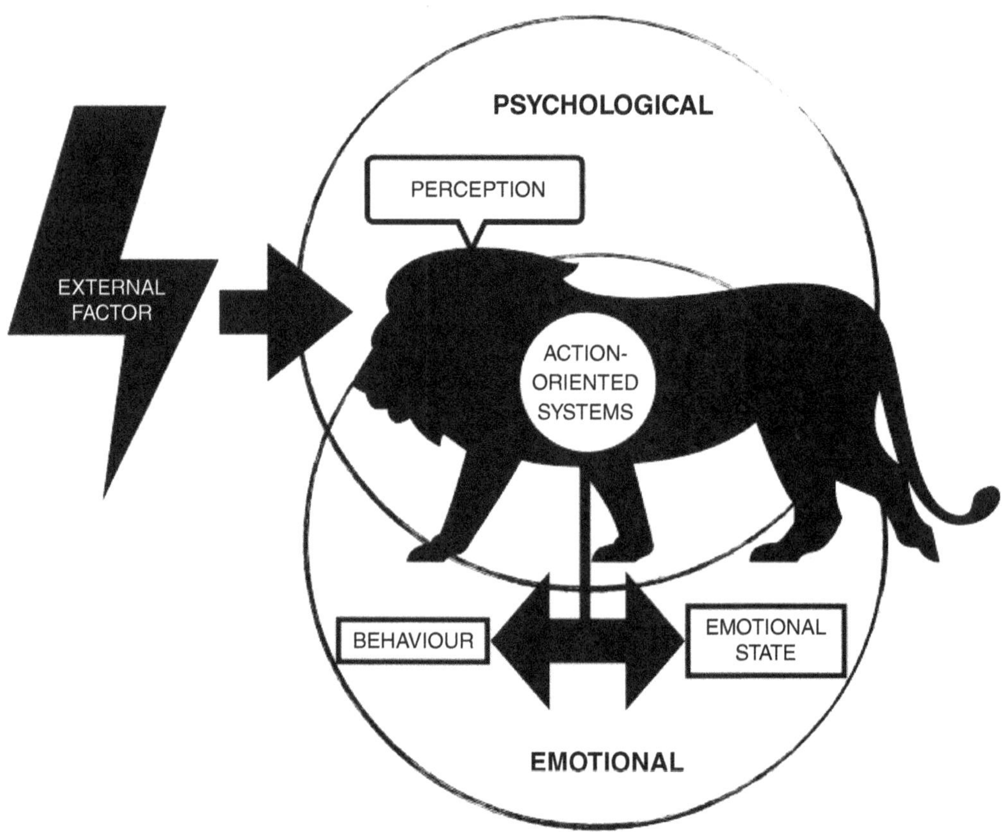

Figure 7.2 Action-oriented systems are the interplay between the psychological and emotional.

- This emotional state helps to sustain the action-oriented response.
  - For example, the emotional state of fear sustains the fight or flight action response and allows the individual to act appropriately in the face of danger. Conversely, a perception that is positive can initiate an emotional response of well-being leading to the action-oriented response of staying near the positive perception, thus sustaining the period of well-being.

Our understanding of such a neurological process comes from a psychobiologist named Jaak Panksepp. He began his study of animals by conducting research on the neural basis of emotions in mammals. He used electrical stimulation of the brain to elicit emotional behaviours in rats and demonstrated that they subjectively experienced these emotions. Through his studies of the neural basis of emotions in animals, he coined the term **affective neuroscience** or the field of research that focuses on studying the neural basis of emotions in both humans and animals. It explores the brain mechanisms and processes involved in emotional experiences, aiming to understand the fundamental nature of emotions and their impact on behaviour and mental health. He identified seven primary emotional systems, including SEEKING, CARE,

PLAY, and LUST, and his work contributed to our understanding of emotions in both humans and animals. These are what Pansepp describes as action-oriented systems. They are controlled by different parts of our brain. This plays a crucial role in understanding emotions and, in humans, developing treatments for emotional imbalances. In animal care, Panksepp's work allows us to understand how these systems can affect animal welfare.[7]

The Neurological Process

The process, by which an external factor is perceived by an individual, initiates an action-oriented response, and determines a behaviour and emotional state, is the **neurological process**. The effects of this process are highly individualised – subject to both the history of the individual and the physiology of the individual.

- *The effects of history*
    The effects of history are highly significant on the entire neurological process. What has been experienced, what has been learnt, what has been given, and what has been deprived have massive effects on the ultimate outcome of an individual's neurological process.
    - *What has been experienced.*
        The source material for everything that is recalled through an individual's memory, are the experiences that an individual has collected over the course of a lifetime. Though experiences may be shared with others, or similar to other's experiences, each individual has a unique perspective on that experience. This unique perspective is what populates an individual's memory. That memory can then be recalled when a similar experience, or a similar external variable, is encountered. This memory allows the individual to immediately judge the perception and act according to this judgement.
        Imagine a sanctuary full of chimpanzees. These chimpanzees may come from very disparate backgrounds. Some may have been kept as pets. They may have been well cared for, but without any other interactions with chimpanzees. Others may have been in laboratories with other chimpanzees. However, they may have had some traumatic experiences. Still, others may have been used in entertainment. Each of these chimpanzees are going to have extremely different experiences, and thus, extremely different memories. Therefore, each will have their own unique perception that is attached to any given external variable. These perceptions will cause the chimpanzees to act in a unique way. In this example, we see why unique care, and a unique understanding for each individual is critical in a managed care
    - *What has been learnt.*
        Individuals *learn* from their unique experiences and attach what they've learnt onto perceptions of external variables. Imagine a squirrel who eats the seed out

---

[7] See Davis & Motag, 2019.

of a bird feeder. Each day, the squirrel depletes the bird feeder of all of its seed; thus frustrating the owner of the bird feeder (who wishes to have birds in his backyard, rather than a few glutinous squirrels). The owner decides to find bird seed that squirrels do not like. He puts this seed in the bird feeder in hopes that the squirrels no longer go to it. Sure enough, it works! The squirrels approach the bird feeder, eat the seed, realise they do not like it, and do not return. The squirrels now have an ingrained memory that they can recall when they see that particular seed in the bird feeder. And when they perceive that seed, they know that they don't like the bird seed, and, likewise, do not approach it.

However, this is not the end of the owner of the bird feeder's frustrations. Not only has he rid the backyard of squirrels, but now the birds are no longer coming to the feeder. He surmises that the new bird seed looks different or smells different and, as a result, doesn't draw the birds in. He decides to mix the old bird seed with the new bird seed in hopes that it attracts the birds. The plan works for a few days.

However, one enterprising squirrel approaches the bird feeder, begins to dig through it, finds that there's some of his preferred bird seed within the new bird seed, and proceeds to claw through all of the seeds, throwing it all on the ground, until he finds the morsels he likes. The other squirrels watch. Now, anytime the owner of the bird feeder fills the feeder with the mixture of seeds, the squirrels know exactly what to do from what they have *learnt*. They empty the bird seeder immediately, finding their preferred seed.

The situation has grown better for the squirrels (and much worse for the owner of the bird feeder). This is all due to the squirrel's ability to *learn*, recall what was learnt, and attach that to the perception of the external variable of the bird seed in the bird feeder.

○ *What has been given.*

When an individual goes from infancy to maturity, they have specific requirements in the course of their rearing. These requirements are both physiological and psychological. For example, some species may require a certain amount of maternal care as they grow up. Some species may require complex socialisation with a group or population.

The degree, quality, and type of what is needed for an animal during the course of their upbringing is going to largely depend of their **life history traits**, or the natural and species-specific stages of an individual's lifetime.

For example, the amount of maternal care is going to vary from species to species – and it is going to be correlated with the level of development an infant has at birth.[8] For example, comparing a giraffe to a chimpanzee shows vastly different life history traits and, therefore, vastly different needs.

Let's look at these differences. Let's imagine that a chimpanzee and a giraffe are conceived on the same day (Figure 7.3). A giraffe gestates for about 15 months. When he is born, the giraffe is basically a small giraffe. He stands up. He looks around, sees the world, and then interacts with his environment in

---

[8] See Thompson et al. 2010.

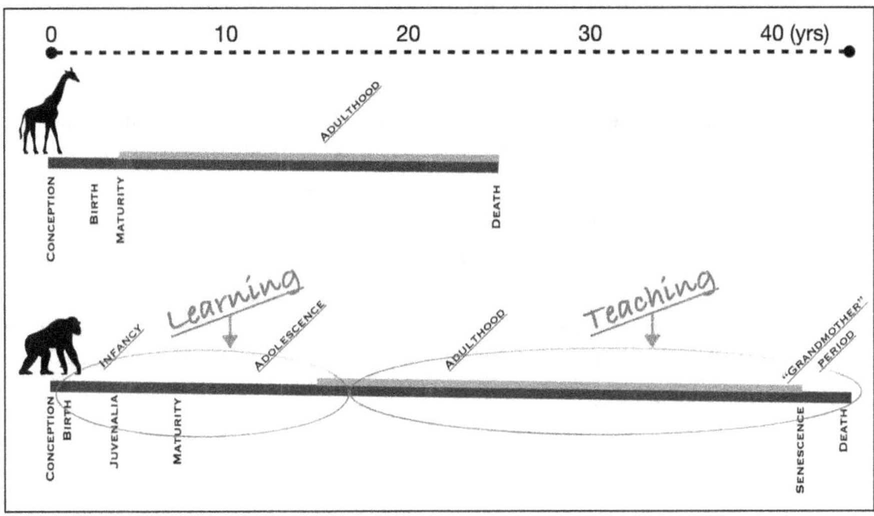

Figure 7.3 Contrasting the life history traits of a giraffe with a chimpanzee.

relatively the way that a giraffe would. He obtains his food by nursing on his mother. However, outside of that, he's pretty much just a giraffe. A chimpanzee, on the other hand, only gestates for about seven and a half months. When the chimpanzee is born, he's completely helpless. He uses his mother for absolutely everything. He clings to her front. He uses her as his limbs, his method of feeding, his protection, and his means of locomotion.

About 4 years into his life, the giraffe becomes sexually mature. It usually takes him a couple of years before he starts actively breeding. But, at that point, the giraffe is a complete adult giraffe. The chimpanzee, on the other hand, has quite a lot of stages go through before this occurs.

For example, chimpanzees go from an infancy period where they cling to their mother's front, are constantly carried around by them, and very dependent on them, to a juvenile period where they typically ride around on their mother's backs, but are more independent. They break away from their mother quite often. They play with other chimpanzees. After about 7 years, chimpanzees reach sexual maturity. However, once they reach sexual maturity, they are still not what one would consider a complete adult chimpanzee.

In fact, they enter a phase called adolescence where they may be sexually mature, but they're still quite dependent on their mother. They run to their mother for protection. Sometimes they sleep near their mother. They eat with their mother. In fact, it's not until typically they reach age 14 before they begin acting like adult chimpanzees. At this point they start siring offspring and are independent of maternal care.

Giraffes live their life as an adult giraffe until their death, usually about 20 after about 25 years. Chimpanzees live quite a bit longer and have an adulthood of about 30 years.

Something very interesting then happens towards the end of a chimpanzee's life. Female chimpanzees enter into a period of reproductive senescence where they can no longer sire offspring, but they're still alive. Some scientists have referred to this as a 'grandmother period', whereby chimpanzees are still very useful to their population and help with younger chimpanzees.[9]

These distinctions are all very important because they highlight how life history traits dictate the species-specifics of what is required in one's own individual history. A chimpanzee survives the world around them by learning. If they are deprived of the various avenues for learning, dictates by their life history traits, it can hamper both their physiological well-being as well as their psychological well-being. In this way, we see how what is given in one's history can be much different between species.

Understanding a species' life history traits can give us a clear idea of the specifics of what is required to have been given to an individual over the course of their lifetime. Life history traits set the historic welfare requirements needed for each individual. What has been given throughout the individual's lifetime, determines the tools they have to interact with the environment around them in a manner that promotes positive well-being.

o *What has been deprived.*

While life history traits allow us to understand what is required for each individual to have proper welfare throughout their lifetime. Life history traits also help us understand the cost of what has been deprived for an individual. Understanding what has been deprived of an individual is critical for a care technician to know in order to provide for individual care needs.

In our example of a chimpanzee versus a giraffe, we can see that, while both individuals require socialisation and maternal care, the costs are dramatically greater for a chimpanzee who has been deprived of maternal care or deprived of a social environment.

Historically, for each individual, the deprivation of species-specific needs can have a profound effect on not just an individual's present state of welfare, but all future states of welfare. These needs can be represented in the form of **affect**, which is the emotional experience associated with the fulfilment or deprivation of a need.

When an animal's needs are not met, it can lead to a negative emotional state, such as hunger, loneliness, fear, or cold. These *affective states* drive the individual to fulfil their needs. The satisfaction of an animal's needs is essential for the balance between their physical needs and the actions created by their emotional state. When their needs are met, it lessens the unpleasant affect and reduces the threat to their life or well-being. On the other hand, when their needs are not fulfilled, the brain continues to deliver unpleasant affect to the animal.

---

[9] See Hawkes 2003.

These emotional experiences can imprint themselves on an animal's action-oriented responses and lead to permanent or semi-permanent negative states of welfare. This occurs when an animal has been conditioned to respond in a manner consistent with the deprivation of needs. Individuals may respond in this manner, even when their needs are present. This can have profound implications for managing the care of an individual.

• *The Effects of Physiology*

An animal's physiology directly influences their ability to engage in action-oriented responses. Positive physiology means that an individual is more likely to exhibit proactive behaviour with regard to interacting with their environment. Conversely, when an animal is suffering from physical ailments, such as injury or disease, their action-oriented responses may be hindered or altered. For instance, an injured animal may exhibit reduced mobility or show signs of pain, affecting their ability to engage in normal behaviours.

Environmental factors, affecting the physiology of an individual, also play a significant role in shaping an animal's action-oriented responses. Temperature, space availability, and the presence of stimuli can all impact an animal's behaviour. For example, an animal exposed to extreme temperatures is physically uncomfortable and may seek shelter or adjust activity levels accordingly. Similarly, an animal confined in a small may be affected physically which can lead to altered action-oriented responses.

Determining the impact of an animal's physical condition on their action-oriented responses requires a **biological functioning perspective**, focusing on changes in physiological responses or behavioural patterns. This perspective considers the animal's biological fitness, immune competence, and physical harmony with their surroundings. Basically, if an animal is acting different from their individual baseline behaviour at the onset of a new physical condition, or if an animal with a chronic physical condition is acting significantly different from the baseline of their conspecifics in the same environment, an impact can be assessed.[10]

Recognising the influence of an animal's physical condition on action-oriented responses is vital in animal care. Uncontrolled pain or stress in animals can lead to compromised health, slower healing, and increased susceptibility to infections. Moreover, animals experiencing distress may exhibit altered behaviour, affecting the outcomes of behavioural assessments. Therefore, considering an animal's physical condition and its impact on action-oriented responses is crucial for ensuring the welfare of an animal in managed care.[11]

Good physical health is essential for optimal animal welfare, enabling animals to engage in natural behaviours and respond appropriately to their environment. Understanding the relationship between physical condition and action-oriented responses is crucial in various contexts in animal care.

---

[10] See Barnett & Hemsworth 1990; and Broom 1991.
[11] See Moberg 2000.

7.3   Assessing and Monitoring Psychological and Emotional Health

Benchmarking

In previous chapters, we have discussed utilising baselines for individuals. Remember that baselines do not mean satisfactory. Instead, baselines merely mean that this is an individual's typical state of affairs, whether behavioural or physiological. Utilising baselines becomes particularly important when assessing psychological health. Each individual has their own psychological proclivities, which are not necessarily positive or negative. Instead, they are their identifiers.

So to utilise universal indicators for all individuals is neither effective, or even realistic. For example, some animals may not be typically as interactive with their environment as others. It would be impossible to divine any useful information about this individual's current state of welfare by using the same tools that measure an individual who is extremely interactive with their environment due to individual preference.

However, there are times and circumstances where utilising individual baselines is not possible. In some cases, you may not have the information at hand to reliably define an individual's baseline. In other cases, you may be interested in comparing the welfare of an entire population. In such cases, a care technician may wish to employ benchmarking. **Benchmarking** is a valuable methodological approach used to assess animal welfare. It involves comparing the welfare of animals in a particular setting to a carefully chosen reference population. This approach provides an objective way to evaluate the welfare of animals and identify areas for improvement. The use of appropriate statistical tests and transparency is crucial in ensuring the validity of benchmarking results.

An example of a study that utilised benchmarking to assess animal welfare is the research on zoo elephants conducted by Clubb and Mason in their review of zoo elephant welfare across European zoos.[12] The study aimed to determine if zoo elephants had substandard welfare and to identify variables that may affect their welfare, such as obesity. The researchers compared the welfare of zoo elephants to benchmark standards and found that improvements were needed to bring zoo elephant performance up to par. This study highlights the importance of benchmarking in identifying areas where husbandry improvements are necessary.

Benchmarking requires careful consideration of the reference population chosen for comparison. The reference population should be representative of the desired welfare standards and should be selected based on relevant criteria. This ensures that the benchmarking results accurately reflect the welfare status of the animals being assessed.

Transparency and judgement are essential when conducting benchmarking studies. It is important to clearly define the criteria used to assess welfare and to document the methods and data used in the study. This allows for transparency and reproducibility, enabling other researchers to evaluate the study's findings.

---

[12] See Clubb & Mason, 2002.

Appropriate statistical tests should be employed to analyse the benchmarking data. Statistical analysis helps determine if the differences observed between the study population and the reference population are statistically significant. This ensures that the benchmarking results are reliable and meaningful.

The Importance of Psychological and Emotional Assessment

As with all aspects of animal health and welfare, assessing and monitoring the psychological and emotional health of animals living in managed care is crucial for ensuring their well-being and quality of life. It allows caregivers to identify and address any deficits or issues impacting general welfare. Objective measures for psychological health, based on observable behaviours and physiological indicators, must be developed to evaluate the animals' well-being. When possible, individual differences among animals within a population must also be taken into account. Regular assessments help caregivers improve the animals' behavioural health and provide positive goals for their care. By monitoring psychological and emotional health, caregivers can identify areas for improvement, prevent potential problems, and enhance the overall well-being of animals in artificial environments.

Let's again look at elephants. Psychological and emotional problems can be common among elephants in captivity and can have significant impacts on their well-being. These problems include stereotypic behaviours, general stress, aggression-related issues, and potential inadequate socialisation. It is crucial to address these psychological challenges to ensure the mental health and welfare of captive elephants. By developing objective measures for each criterion of psychological and emotional health, care technicians can effectively evaluate and address the need and psychological well-being of elephants. Without objective criteria and regular assessments, care technicians might only intervene when elephants are visibly suffering. By having measurable criteria for psychological and emotional health, care technicians can proactively improve the well-being of elephants and prevent potential problems from escalating.

As with all other animals, in order to develop objective measures for psychological and emotional health, it is essential to consider what can be objectively observed in elephants – both behaviourally and physiologically. Observable behaviours, such as baseline interactiveness of an elephant towards humans (care technicians and/or the public), can provide valuable insights into the psychological state of elephants. Additionally, physiological measures, such as immune system functioning, can offer further evidence of their well-being. By combining these measures, categories can be defined to identify low, normal, and high functioning, similar to how physical wellness is assessed through blood tests.[13]

Elephants represent just one example. We choose elephants as our example because their psychological and emotional needs can be so clearly observed. However, all animals in managed care must be assessed for their psychological and

---

[13] See Mason & Veasy, 2010.

emotional well-being. These assessments are going to vary from species to species in terms of which tool is the most effective. For example, the psychological and emotional assessment of a chimpanzee is going to differ from that of an elephant. Even more significantly, the psychological and emotional well-being of a snake is going to be extremely different from that of a chimpanzee. It is incumbent on the care technician to research their species and make a determination as to the best assessment tool that they can perform.

Obviously, the importance of individual differences in these assessments is enormous. Therefore each criterion should be measured against an individual's baseline when possible. By recognising and accommodating these individual differences, not just during assessments but also in planning care, care technicians can provide tailored care and support to promote the well-being of each animal.

7.4   Care Plans

The crux of individualised care rests in the actualisation of a written individualised care plan. Along with individual care plans that address physiological or veterinary needs, care plans can also address psychological and emotional needs. After the assessments have been performed, and an animal's psychological and emotional needs are clearly defined, a written individualised care plan is the next step.

A good care plan sets definable and attainable goals, details an explicit strategy for attaining these goals, sets a realistic plan of implantation, specifies a manner of documenting the effects of the plan, details how the effects are to be assessed and evaluated, and sets an expiration date for either renewal, amendment, or cancellation based on the assessment and evaluation.

Setting Goals

The first, and perhaps most critical, step of an individual care plan is to define exactly what goals one is hoping to attain with the plan. For example, if an animal is experiencing undue stress and anxiety in situations that are not normally stressful or anxious, the goal may be to desensitise that individual to whatever is triggering the anxiety. Perhaps an individual gets overly aggressive with conspecifics during a particular time of day. The goal may be to distract that individual in order to minimise the aggression and, over time, have this aggression abate. Defining these goals is critically important because it is against these goals that the effectiveness of the care plan is to be measured.

Detailing a Strategy

A proper care plan very specifically details the methods that will be employed in order to achieve the goals. These methods are the plan's strategies. Strategies need to be as specific and as detailed as possible because, during the assessment and evaluation phase of the care plan, it is the strategies that are able to be amended and tweaked. If strategies are too broad, the plan becomes very difficult to properly amend when it is not as effective as hoped.

Details should include: which personnel; when, during the of day, the plan is to be carried out (and how regularly); where, locationally, the care plan is occurring; and a written justification for each step of the strategy. The justification is also an important piece of the care plan because when it comes time to edit the strategy, the original justification needs to be noted for why a certain aspect of the care plan was done in a certain manner.

Implementation

The implementation of these strategies must be carefully thought out. The act of performing these strategies must be realistic in terms of available resources, staffing, skill level of staff, and safety. The logistics of implementation should be specified directly in the care plan.

The care plan must also be *consistently* implemented by all staff involved. This ensures that the individual receives consistent and appropriate care. Ensuring consistent implementation also helps to rule out inconsistency as a cause for any issues or inconsistencies with the plan.

Documentation

Once the care plan is implemented, it's important to document how the plan is affecting the target individual. Daily ad libitum notes from care technicians should be coupled with standard data sheets that allow the care technician to record predetermined relevant benchmarks, positive, negative, or neutral. These notes need to be reviewed regularly by husbandry and veterinary staff to ensure that the care plan is progressing accordingly.

In addition to keeping notes about the implementation and operation of the specific care plan, care technicians should enhance their routine daily records with regard to it. Daily records enable care technicians the ability to monitor the animal's response to the enacted care plan. By recording the animal's behaviour, interactions, and any changes in its physical condition, care technicians can evaluate the effectiveness of the care plan and make necessary modifications. Tracking the proportion of time the animal spends engaged in positive versus negative activities, such as locomotion, foraging, investigating, and other enriching behaviours, caretakers can evaluate the animal's level of stimulation, satisfaction, and overall welfare.

Documentation can also benefit future care – both at the facility where the care plan is being enacted and other facilities that might be experiencing the same challenges. Long-term records can contribute to scientific studies, help identify trends or patterns in behaviour, and contribute to the understanding of captive animal management and welfare. If this information is retained and shared, it can improve the overall welfare of animals with similar challenges by showing what worked and what didn't work.

Evaluation and Assessment

All of this leads to an overall assessment of the individual care plan. Such an assessment couples the daily records being kept by care technicians with an overall care and welfare assessment method. Other objective measures may also be employed to determine the effectiveness of the care plan. The assessment should look directly at the goals and determine which of those goals are affected, if any. The assessment period should also look at other aspects of the individual's well-being to ensure that the care plan is not actually having a negative effect. Inherent within the goals or a timeline of when expected results should occur. The care plan should be measured against these expected results. The end result, or conclusions, drawn from an assessment should look something like this:

- How did the individual care plan measure against the intended goals?
- If goals were accomplished, were they accomplished in an acceptable time frame?
- Were there unintended consequences of the care plan? Were these consequences negative? Neutral? Positive?
- What about the strategies could be amended to have a greater chance of success? What objective data supports these amendments?

Renew, Edit, or Abandon

All care plans should carry with them a standard expiration date. At the end of this expiration date, as long as the care plan has been implemented according to the plan, the assessment is done and analysed. At that point, the husbandry team must decide if the plan is to be renewed, edited, or abandoned altogether.

Renewal means that the care plan is progressing according to its stated goals. It means that its effectiveness is being shown in the assessment period. Finally it means that more time is needed for it to be fully realised.

If the care plan is to be amended then it is not satisfactorily hitting the stated goals. Therefore, the strategies should be tweaked according to information gleaned from the assessment period, then resumed. At the next expiration date, it will then be assessed once again to see if these amendments have aided the effectiveness of the plan.

If the care plan is to be abandoned altogether, it has either been *completely* successful in hitting its stated goals (and there is no more need for these new protocols), or the plan is actually having no effect at all. In some cases the plan may be having a *negative effect on the* individual and this is not shown until the assessment is completed. At that point, the husbandry team should decide if the plan should be amended or tweaked.

Implications

How we internalise the environment around us, shapes and determines our well-being. How we attach negative, neutral, or positive connotations to every external variable we experience creates our general welfare. This psychological and emotional process determines our perception of the world around us, our emotional state, and our reactions. Psychological and emotional health is, therefore, critical to this welfare.

We may live in the most resource rich, safe, and engaging environment possible; however, if we have negative connotations onto our perceptions of those external variables, it will affect not just our emotional state but also the actions we take in that environment. Conversely, we may be in the most challenging of environments, but if we are able to positively internalise those external variables, we can maintain a more positive general well-being. A mariner facing stormy seas can only survive the event with a productive neurological process – to take those challenging external variables, maintain a calm, emotional state, and then choose appropriate reactions to the challenges brought upon by the situation.

As care technicians, we may be providing an engaging environment. We may be providing all the necessary resources. We may be providing absolute safety. However, in order for these factors to be truly positive for the animals in our care, we must ensure psychological and emotional health – a state of health that can be impacted by past histories, as well as physiology. This understanding for every individual in our care, along with scientific observations of each individual's behaviour, can allow us to understand each individual's psychological and emotional health, and effectively treat any issues. These practices can go a long way in our quest to ensure the opportunity to thrive for each and every individual in our care.

References

Barnett, J. L. and Hemsworth, P. H., 1990. The validity of physiological and behavioural measures of animal welfare. *Applied Animal Behaviour Science*, 25(1–2), pp. 177–187.

Broom, D. M., 1991. Animal welfare: concepts and measurement. *Journal of Animal Science*, 69(10), pp. 4167–4175.

Clubb, R. and Mason, G., 2002. *A review of the welfare of zoo elephants in Europe* (p. 303). Horsham: RSPCA.

Davis, K. L. and Montag, C., 2019. Selected principles of Pankseppian affective neuroscience. *Frontiers in Neuroscience*, 12, p. 1025.

Hawkes, K., 2003. Grandmothers and the evolution of human longevity. *American Journal of Human Biology*, 15(3), pp. 380–400.

Hetts, S., Estep, D. and Marder, A. R., 2005. Psychological well-being in animals. *Mental Health and Well-Being in Animals*, pp. 211–220.

Mandik, P., 2005. Action-oriented representation. *Cognition and the Brain: The Philosophy and Neuroscience Movement*, pp. 284–305.

Maple, T. L., 2015. Four decades of psychological research on zoo animal welfare. *Markus Gusset1 & Gerald Dick2*, 24, p. 41.

Mason, G. J. and Veasey, J. S., 2010. How should the psychological well-being of zoo elephants be objectively investigated?. *Zoo Biology*, 29(2), pp. 237–255.

McMillan, F. D., 2002. Development of a mental wellness program for animals. *Journal of the American Veterinary Medical Association*, 220(7), pp. 965–972.

McMillan, F. D., 2003. Maximizing quality of life in III animals. *Journal of the American Animal Hospital Association*, 39(3), pp. 227–235.

McMillan, F. D., 2005. The concept of quality of life in animals. *Mental Health and Well-being in Animals*, pp. 181–200.

Mellor, D. J., 2012. Animal emotions, behaviour and the promotion of positive welfare states. *New Zealand Veterinary Journal*, 60(1), pp. 1–8.

Moberg, G. P., 2000. Biological response to stress: implications for animal welfare. In Moberg, G. P. and Mench, J. A., *The biology of animal stress: basic principles and implications for animal welfare* (pp. 1–21). Wallingford: CABI Publishing.

Panksepp, J., 2005. Affective consciousness: core emotional feelings in animals and humans. *Consciousness and Cognition*, 14(1), pp. 30–80.

Panksepp, J., 2011. Toward a cross-species neuroscientific understanding of the affective mind: do animals have emotional feelings?. *American Journal of Primatology*, 73(6), pp. 545–561.

Thompson, K. V., Baker, A. J. and Baker, A. M., 2010. Parental care and behavioral development in captive mammals. In Kleiman, D. G., Thompson, K. V. and Baer, C. K., *Wild mammals in captivity: principles and techniques* (pp. 367–385). Chicago: University of Chicago Press.

8   Physiological Health

Figure 8.1 A Warthog at the Kansas City Zoo & Aquarium. (Photo credit: author).

8.1   Liberation

In perfect isolation, the castaway sits on the shore and stares into the fading dusk. He's been alone for so long that he has no real memory of what it's like to interact with another human. The scars he's received, the injuries that have befallen him, and the illnesses that he's faced – all during his time on the island – have all left an indelible mark on his physical being. In fact, he no longer remembers how it feels to live without them. As the sun is eclipsed behind the horizon, and the moon peers through a dark and cloudy sky, the castaway faces yet another night in his solitary home.

At the edge of the horizon, appearing through the night air, a distant light materialises on the water. The castaway catches a glimpse of the light, rubbing his eyes to try to make out what it is. As the wind whistles through the trees behind him, and the waves crash on the shore in front of him, the light becomes nearer. Soon, he can make out that the light is attached to a ship.

Instantly, the castaway is filled with fear. He hasn't interacted with anyone in so long. He doesn't know what harm might befall him by this mysterious ship. He has no thoughts of rescue. He only has the urge to flee. He tries to run away, but his knees give way and he falls onto the sand. He crawls towards the trees. As he struggles, he looks back into the water. The ship is now closer. He can make out a figure, the figure of one mariner.

All hopes of being rescued have faded years ago. The castaway has no thoughts of this possibility. He has grown accustomed to the lonely world. He has grown accustomed to the isolation. He has grown accustomed to the ailments that his body faces every day on this island. He doesn't even recognise them anymore.

Consumed by this abject fear, the castaway buries his face into the sand, covering the back of his head with his hands. He cannot see that the ship has pulled up to the shore. He cannot see that The Mariner has waded through the water and is now standing behind him.

The sounds of another human approaching him are a foreign and distant memory. He recognises not the reality of what these sounds indicate but merely the confusion of the unknown.

'Who are you?', a voice from behind him calls.
The castaway does not answer.
'Where are you from?', asks the voice.
Still the castaway doesn't answer.
'How long have you been here?'
The Castaway tries to bury his face deeper into the sand. The Mariner bends down and grabs the arm of the castaway. The castaway flinches. With great force, The Mariner turns the cast castaway over.
The Mariner sees the sun damaged face of the castaway. His long beard, covered in sand, indicates a man who has been stranded for years, if not decades.
The castaway, keeping his eyes closed, tries to turn back around. The Mariner pins him down.
'Look at me!', says The Mariner.
Slowly the castaway opens his eyes.
The Mariner calms his voice. 'I can help you. I can get you off of this island. I have a ship. I'm on a journey. You can join me.'
The castaway stares at him, confused. Slowly, the castaway sits up. He looks at The Mariner.
The Mariner smiles. 'Let's get you up.'
He extends his hand and attempts to pull the castaway up. Immediately, he notices that the castaway's knees keep buckling under him. He can't straighten his legs. In response, The Mariner throws the castaway's arm over his neck and lifts him up as they walk to the water.

When they get to the shore, The Mariner puts the castaway down.

'Let's walk across the water to the ship. That way, we can get you cleaned up and get you a good meal.'

As they walk through the water, The Mariner has to steady the castaway. Finally, they get to the ship and approach the rope ladder. It becomes very evident that the castaway is not going to be able to make it up the ladder. The Mariner tries to lift him, but the castaway's knees are too weak. Each time, he falls back into the water

The Mariner devises a plan. He climbs the ladder, goes to the deck, and approaches the pulley system that holds the lifeboats. He loosens the rope and drops the lifeboat into the water. It nearly falls on the castaway as it creates an enormous splash.

'Climb on!', yells The Mariner.

The castaway looks confused. Frustrated, The Mariner climbs down the rope. Struggling, he lifts the castaway up and drops him into the lifeboat.

'Stay there', instructs The Mariner.

He climbs back to the deck and hoists the castaway up. When he reaches the deck, he lifts the castaway out of the boat and puts him on the deck.

'Can you walk?', he asks.

The castaway stares at him, but slowly stands up and hobbles across the deck.

'There is a cabin and a place to bathe over there. I will prepare you some food. Can you understand me?'

The castaway nods. He walks over to the cabin, opens the door, and closes it behind him.

An hour passes. The Mariner has prepared a meal in the ship's hold. The castaway appears. He has cleaned himself up. He has even shaved. He mutters out a few words.

'Where am I?', he asks.

'You are on my ship. I found you on this island. Can you tell me how long you have been here and where you come from?', asks The Mariner.

The castaway closes his eyes. 'I do not know. Everything is hazy in my mind. I have been here for a long long time.'

'I want to help you', explains The Mariner. 'But first, I am on a quest. I am on my way to the far north sea. It is said that a very unique type of tree exists in this area and the volcanic ash of an island. I am out to harvest the berries from this tree, which can grant me a small fortune. I will use this money as part of my redemption for sins that you can't even imagine. Once I have found this tree and harvested the berries, I will bring you home. Perhaps in the time it takes us to find the tree, you'll regain your memories and I can bring you to where you are from.'

The castaway nods, seemingly understanding what The Mariner has told him.

The Mariner continues, 'Clearly you have very serious physical challenges. You can't walk. Your skin is burnt. You have had poor nutrition. I need to know what is wrong with you so that I can help you take this journey.'

The castaway, having lived with these maladies for so long, shakes his head. He is unable to communicate anything about his physical condition.

The Mariner places his hand on the castaway's shoulder. 'It's okay. We will figure this out.'

When Needs Cannot Be Communicated

What do you do when the individual you're caring for cannot let you know what's wrong? How can you provide individual care if you don't know what one's individual challenges and needs are?

In our story, the castaway cannot relate anything to his rescuer. He cannot tell him anything about his history. He cannot tell him what he's been through. He cannot even tell him what hurts. So it's up to The Mariner to observe him. The Mariner has to ascertain what the castaway can do and what he can't do. Then it's up to The Mariner to figure out ways to abate these challenges.

Firstly, The Mariner has to win the castaways' trust. The castaway has to be comfortable enough to act and move about around The Mariner. If not, all the castaway is going to do is act out of fear. In turn, all The Mariner is going to observe is the castaway fleeing or fighting. This is not going to give him a true understanding of the castaway's physical challenges. However, if the castaway trusts that he is not in danger from The Mariner, he can begin to act as he normally would. This will give The Mariner a true indication of what is occurring with him.

It becomes plainly obvious that the castaway cannot walk well. He is clearly going to need assistance. It becomes even more evident that the castaway is not going to be able to get up the ladder.

So, in order for The Mariner to get the castaway to a space where even the most minimal care can be achieved, he has to devise a novel solution where he lifts the castaway up to the ship's deck by use of a pulley system and a lifeboat.

In the care of animals, it is up to us as care technicians to assess what those needs and challenges are. We've already discussed emotional and psychological needs. However, divining these physical needs takes an even more meticulous approach.

8.2   Defining Physiological Health

Physiological health refers to the state of an animal's body and its ability to function optimally. It encompasses various aspects, including physical well-being, disease resistance, and the absence of physical discomfort or pain. In the context of animal welfare, physiological health is a crucial component as it directly impacts an animal's overall quality of life. Physiological health is central to animal welfare. In its simplest context, positive physiological health contributes to positive welfare, while negative physiological health can lead to poor animal welfare. When animals are in good physical health, they experience a state of well-being that allows them to engage in natural behaviours, experience positive emotions, and adapt appropriately to their environment.

Animals living in artificial environments such as zoos, sanctuaries, or research facilities face a different set of variables in terms of physiological health than animals in the wild. While these settings aim to provide for the animals' basic needs, there are inherent challenges that can affect their physiological health. As we have discussed, these challenges include limited space, restricted movement, altered social dynamics, and exposure to stressors that do not exist within their natural habitats (Broom, 2001).

Other challenges can include the increased risk of disease transmission due to close proximity with both conspecifics and other species (including humans) and limited natural barriers between them – leading to higher rates of infection and compromised immune systems. Limited space or limited stimulation can lead to a lack of physical exercise, which can result in reduced muscle tone, obesity, and other physical ailments.

Wild animals, on the other hand, live in their natural habitats and are subject to a range of environmental factors that can impact their physiological health. In the wild, animals have access to a diverse range of food sources, natural shelters, and opportunities for physical activity. These factors contribute to their overall well-being and help maintain their physiological health.

However, wild animals also face challenges such as predation, competition for resources, and exposure to harsh environmental conditions. These factors can lead to injuries, diseases, and other physiological stressors. Nevertheless, wild animals have evolved over time to adapt to their environments, developing natural defence mechanisms and behaviours that promote their physiological health.

None of this is to suggest that the challenges faced by animals in managed care cannot be mitigated. Rather it is to emphasise how critical it is to recognise the different challenges faced by animals in artificial environments. By recognising these challenges, we can implement measures to address them in order to give every animal in our care the means to physiologically thrive.

This requires us to recognise what constitutes physiological thriving. Overall, we can say that physiological thriving means the state of feeling free of physical malady coupled with the ability to physically perform all natural behaviours and inclinations. For example, if we are an arboreal species, we need to be physically able to climb and move about through a canopy. If we are a nomadic species, we need the physical stamina to traverse long distances. If we are a predatory species, we need the physical ability to stalk and run after prey.

We can summarise this in the following manner. Physiological health requires:

- *Freedom from disease and injury*: Animals that are free from disease and injury experience reduced pain, discomfort, and suffering. They can engage in normal activities, exhibit natural behaviours, and maintain a good quality of life.
- *Adequate nutrition and hydration*: Providing animals with proper nutrition and access to clean water ensures their physiological needs are met. This promotes overall health, vitality, and the ability to thrive in their environment.
- *Suitable environmental conditions*: Animals require appropriate temperature, humidity, and ventilation to maintain physiological homeostasis. When these conditions are met, animals can regulate their body temperature, breathe comfortably, and avoid stress-related health issues.
- *Sufficient space and exercise*: Satisfying an animal's need for space and exercise promotes physical fitness, muscle development, and overall well-being. It allows animals to engage in natural behaviours, prevent obesity, and maintain healthy cardiovascular and musculoskeletal systems.

Conversely, when animals experience negative physiological health, it can have detrimental effects on their welfare. We can summarise how negative physiological health contributes to poor animal welfare in the following manner:

- *Disease and injury*: Animals suffering from diseases or injuries may experience pain, discomfort, and reduced mobility. This can limit their ability to engage in natural behaviours, negatively impacting their overall welfare.

- *Malnutrition and dehydration*: Inadequate nutrition and lack of access to clean water can lead to malnutrition, dehydration, and weakened immune systems. These conditions can result in poor health, reduced vitality, and increased susceptibility to diseases.
- *Inadequate environmental conditions*: Exposure to extreme temperatures, poor air quality, or overcrowded conditions can cause stress, respiratory problems, and compromised immune systems. Such conditions can lead to chronic health issues and decreased welfare.
- *Lack of exercise and enrichment*: Animals deprived of sufficient space, exercise, and mental stimulation may develop physical and behavioural problems. This can include obesity, muscle atrophy, stereotypic behaviours, and reduced overall well-being.

In short, positive physiological health contributes to positive welfare by allowing animals to thrive, engage in natural behaviours, and experience overall well-being. Conversely, negative physiological health can lead to poor animal welfare, resulting in pain, discomfort, and compromised physical and mental states. By prioritising and promoting positive physiological health, we can enhance the welfare of animals and ensure their overall well-being.

8.3   Assessing and Monitoring Physiological Health

The importance of consistently monitoring and assessing an animal's physiological condition cannot be understated (Mason & Mendl, 1993). What is paramount in these assessments is the ability to get objective and valid data on an individual's general condition. As with monitoring psychological and emotional health, monitoring physiological health requires methods and protocols with the ability to draw valid conclusions from objective data. However, unlike psychological and emotional assessments, there are times where physiological assessments come with a degree of risk. At times, this risk can be extremely significant. Risky physiological assessments are usually **invasive techniques,** or those that require penetrating the body. This is contrasted with **non-invasive techniques**, which are largely observational or can be performed without delving into the body of an individual.

Invasive techniques can range from something as simple as blood sampling, all the way to performing surgical biopsies while an animal is under anaesthesia. These techniques generally offer extremely precise data on an animal's health status. However, due to the risk, these methods are often reserved for situations where detailed diagnostics are necessary and non-invasive alternatives cannot provide the requisite depth of formation. They're particularly useful in diagnosing complex conditions or when monitoring the progression of a disease.

Conversely, non-invasive methods offer a fairly risk-free ability to gain insight into an animal's health status. Non-invasive methods such as faecal analysis, the use of scales, body condition scoring, or even the use of wearable technology to monitor heart rates can oftentimes give the care technician critical information without the stress or risk of an invasive procedure.

The choice between invasive and non-invasive methods can be a complicated one. The choice generally hinges on a careful consideration of the animal's welfare, the specific information needed, and the context of the assessment. It is often a choice that must take into consideration the delicate balance between obtaining critical health data and minimising risk and stress to an individual.

Non-invasive Methods

Non-invasive techniques are generally pain-free, stress-free, and have minimal, if any risk to the individual, they can be performed on a regular basis, thus allowing a care technician to consistently assess animal health. These methods can be as simple as strategic observations of an animal's movement, or as complicated as utilising operant conditioning to perform imaging procedures on an individual. In some cases, non-invasive techniques can give a similarly detailed assessment of an animal's physiological health. For example, faecal and urine analysis provide valuable insights into an animal's metabolic and digestive health. Utilising operant conditioning to train individuals on the use of a scale can provide precise weight measurements. Some individuals can even be trained to receive an echocardiogram for a heart condition.

Below are some examples of non-invasive techniques that are commonly used to provide insight into an animal's health.

* *Body condition scoring:*

    Body condition scoring (BCS) is a non-invasive, yet highly informative method used to assess the general health and physical well-being of animals. This technique involves evaluating the animal's fat and muscle mass visually and through palpation, assigning a score based on a standardised scale. Regular application of BCS is crucial for monitoring health, with assessments recommended to be performed annually for each individual. However, more frequent evaluations may be warranted based on changes in health status or as part of routine health checks.

    The data collected from BCS assessments serve multiple purposes. Firstly, it provides a baseline for individual health, allowing for the detection of any deviations from an animal's normal condition. This can be instrumental in early identification of health issues, enabling timely intervention. Secondly, the aggregated data over time can reveal trends in the health status of the population, guiding adjustments in diet, husbandry practices, and veterinary care. Finally, this data should be meticulously recorded and integrated into the animal's health records. Regular reviews of these records by veterinary and care staff can inform the development or adjustment of individualised care plans, ensuring that each animal receives the most appropriate and effective care based on their specific needs and condition.

    BCS techniques can involve gathering an array of information, oftentimes based on the species of the information prioritised. The following is a standard template:

    Scoring criteria (1–5):
    General appearance (e.g., posture, gait)

Visible fat deposits (e.g., around the abdomen, neck)
Muscle mass visibility (e.g., limbs, back)
Overall body condition score:

Scoring should use the following metrics::
1 = Severely underweight
2 = Underweight
3 = Ideal
4 = Overweight
5 = Obese
Additional Notes:

- *Quality of life assessments*

  Quality of life assessments (QoL) are prescribed physiological assessments performed by care technicians and veterinarians to determine the overall physical quality of life through factors such as mobility, signs of pain, BCS, dietary consumption, and daily behaviours. These assessments are vital as they provide insights into the animals' health, comfort, and overall satisfaction with their living conditions. The importance of QoL assessments lies in their ability to inform caregivers about the needs of the animals, helping to make informed decisions regarding their care, including necessary adjustments to their environment, diet, and social interactions.

  QoL assessments involve utilising a template to observe an animal for signs of stress or discomfort, evaluating its physical condition for any signs of illness or injury, and assessing behaviours. Templates are critical for allowing care technicians to take consistent and objective data, providing a structured approach to evaluating the animal's well-being.

  QoL assessments are utilised both routinely – for identifying potential issues early on, as well as in a prescribed manner – such as when an animal is extremely ill or when euthanasia is being considered.

  QoL templates can differ from facility to facility and may be species-specific. An example of a general template looks like this[1]:
- General Information
  - Facility Name:
  - Date of Assessment:
  - Assessor(s):
  - Animal Information:
  - Species:
  - Name/ID:
  - Age:
  - Sex:
- Physical Health:
  - Visible Health Issues (Yes/No):
  - Details of Health Issues:

---

[1] See Campbell-Ward, 2023.

- o Mobility (Good/Average/Poor):
- o Signs of Pain or Discomfort (Yes/No):
- o Body Condition Score:
- Behavioural Well-being:
  - o Engagement in Natural Behaviours (Frequent/Occasional/Rare):
  - o Signs of Stress or Anxiety (Yes/No):
  - o Interaction with Enrichment (High/Medium/Low):
- Social Environment:
  - o Interaction with Conspecifics (Positive/Neutral/Negative):
- Overall QoL Rating:
  - o (Excellent/Good/Fair/Poor):
- Assessor's Notes and Recommendations:

- *Utilising operant conditioning*

  In Chapter 9, we will explore the use of operant conditioning, or animal training, for a variety of purposes. Chief among these is utilising operant conditioning for veterinary care. This approach leverages the principles of positive reinforcement to train animals to voluntarily participate in their own health assessments, thereby reducing stress and the need for restraint or sedation. For instance, animals can be trained to stand on a scale, stand still for ultrasound examinations, or open their mouths for dental checks. The process involves breaking down the desired behaviour into small, manageable steps and rewarding the animal for each step successfully completed, gradually shaping the desired outcome. This method not only facilitates easier and safer medical examinations but also enhances the welfare of the animals by incorporating mental stimulation and trust-building into their routine care.

Invasive Methods

Though invasive methods of assessing and monitoring physiological conditions can come with stress to the animal and various levels of risk to the animal's safety, sometimes they are unavoidable. They give precise information, which is oftentimes critical to being able to treat a physiological challenge. They offer detailed insights that non-invasive methods may not provide. These techniques, while more intrusive, can yield invaluable data on an animal's health status, particularly in diagnosing or managing complex conditions.

Invasive methods of assessing and monitoring an animal's physiological well-being should only ever be performed under the guidance of a veterinarian. Some examples of invasive methods that a veterinarian might choose to undertake are below.

- *Blood sampling*

  Blood sampling in animals is a fundamental diagnostic tool in veterinary medicine, offering insights into the systemic health of an animal. This procedure involves collecting a small amount of blood, usually from a vein, to perform various tests. The process requires skill and care to minimise stress and discomfort to the animal.

The site of blood collection is often shaved and cleaned to prevent contamination. A trained veterinarian or veterinary technician usually performs the procedure, using sterile equipment to ensure the safety and well-being of the animal.

The collected blood can be analysed for a complete blood count (CBC), which provides information on the number and type of blood cells, indicating conditions such as anaemia, infection, or inflammation. Biochemical profiles assess organ function and electrolyte status, crucial for diagnosing diseases affecting the liver, kidneys, or metabolic disorders. Blood samples can also be used for more specialised tests, including hormone levels, infectious diseases, and genetic testing.

- *Biopsies*

  Biopsies is a surgical process for the examination of tissue samples from living animals to diagnose diseases, including cancer. This invasive method can be performed on various tissues, including skin, internal organs, or bone marrow, depending on the suspected condition.

  There are several types of biopsies, including needle aspiration, punch biopsy, incisional biopsy, and excisional biopsy. Needle aspiration involves using a fine needle to collect cells from lumps or organs, suitable for superficial masses or for organs like the liver or kidney. Punch biopsies, often used for skin lesions, involve using a circular blade to remove a small cylinder of tissue. Incisional biopsies remove a small portion of a larger mass, while excisional biopsies aim to remove an entire lump or area of interest, providing a comprehensive sample for analysis.

  The procedure requires careful planning, including the selection of the biopsy site and the determination of the most appropriate type of biopsy. Sedation or anaesthesia is commonly used to ensure the animal's comfort, minimising stress and pain.

  After the biopsy, the sample is sent to a veterinary pathologist for microscopic examination. The results can provide vital information about the nature of the disease, guiding treatment decisions and helping to predict the animal's prognosis.

- *Endoscopy*

  Endoscopy is a minimally invasive diagnostic tool widely used by veterinarians to monitor and assess the physiological well-being of animals. This technique involves the use of an endoscope, a small camera that is attached to a flexible tube, allowing for the visual examination of an animal's internal organs and surfaces without the need for extensive surgery. Endoscopy can be applied to various parts of the body, including the gastrointestinal tract (gastroscopy), respiratory system (bronchoscopy), and abdominal cavity (laparoscopy).

  The primary advantage of endoscopy is its ability to provide real-time, detailed images of the internal structures, enabling veterinarians to identify abnormalities such as ulcers, tumours, and foreign objects. It also allows for biopsy procedures, where small tissue samples can be collected for further histological examination, aiding in the accurate diagnosis of diseases.

Endoscopy is particularly beneficial as it significantly reduces the need for more invasive surgical explorations, thereby minimising the risk of complications and promoting quicker recovery times for the animal.

- *Surgical exploration*

Surgical exploration is a more invasive method, used when other diagnostic tools are inconclusive. It allows veterinarians to directly observe the condition of internal organs and tissues, take biopsies, and sometimes resolve the issue during the same procedure.

8.4   Providing Evidence-Based Physiological Care

As with providing for optimal psychological and emotional health, providing for optimal physiological health demands an evidence-based approach. Gleaned from both general data about the species, and individual data from in-house assessments, care technicians can provide care that ensures that animals live in a manner that promotes positive physiological well-being.

Nutrition

A robust nutrition program is fundamental to the overall health and well-being of animals, particularly in settings such as sanctuaries and zoos. The significance of such a program cannot be overstated, as it directly impacts the animals' physical health, psychological well-being, and longevity. Tailored dietary plans, based on individual needs through physiological assessments, ensure that each animal receives the optimal balance of nutrients required for their specific health needs, life stage, and any medical conditions they may have, such as diabetes or kidney disease.

Generally, a nutrition program is meticulously designed by a team comprising a veterinary nutritionist specialising in animal nutrition, along with care technicians, and possibly a commissary technician. This collaborative approach guarantees that the diets not only meet the nutritional needs of the animals but also consider their feeding ecology and individual health status. High-quality food items are selected to mimic the animals' natural diet as closely as possible, promoting natural foraging behaviours and enhancing their quality of life.

Moreover, the program's effectiveness is continuously evaluated and adjusted based on the ongoing health assessments of the animals, including body condition scoring and monitoring for any health issues that may arise. This proactive and individualised approach to nutrition underscores its critical role in animal care, ensuring that each animal thrives in their care environment.

However, while the nutritious content of an animal's diet is of paramount importance, the diet must still be both enriching and appetising. Crafting a diet that is both nutritious and enriching requires a thoughtful balance between meeting the physiological needs, stimulating the natural behaviours of the animals, and finding preferred foods (both on a species level and an individual level). Beyond mere nutrition, the presentation and variety of the diet play a critical role in ensuring that individual's actually consume their diet. It also promotes mental stimulation and natural foraging behaviours.

To achieve this balance, facilities can employ several strategies. Firstly, incorporating a variety of food items that mimic the animal's natural diet can encourage exploration and problem-solving. For example, hiding food within enclosures or using puzzle feeders can simulate the challenges animals face in the wild when searching for food, providing both mental and physical engagement.

Secondly, the timing and method of food delivery can be varied to prevent predictability, further enriching the animals' daily routine. This unpredictability mirrors the natural unpredictability of food availability in the wild, keeping the animals mentally alert and engaged.

Lastly, involving animal care staff in the creative process of diet presentation can lead to innovative feeding methods that enrich the animals' lives. By balancing the nutritional content with enriching presentation methods, facilities can ensure their animals lead physically healthy and mentally stimulating lives.

Nutrition programs should always incorporate written protocols. An example of a nutrition protocols looks like this:

- Introduction
  - Brief overview of the program's purpose: To ensure all animals receive a balanced, species-appropriate diet that supports their health, well-being, and natural behaviours.
- Nutritional requirements
  - Outline of species-specific dietary needs based on age, health status, and activity level.
  - Consideration of special needs for pregnant, nursing, or geriatric animals.
  - Calorie budgets.
- Food selection
  - List of approved food items for each species, emphasising variety and quality.
  - Guidelines for sourcing and storing food to maintain freshness and nutritional value.
- Feeding strategies
  - Schedule: Frequency and timing of feedings to mimic natural feeding patterns.
  - Presentation: Techniques to encourage natural foraging and hunting behaviours, such as scatter feeding, puzzle feeders, and hidden food.
- Health monitoring and adjustments
  - Regular body condition scoring and health assessments to monitor the impact of diet.
  - Process for adjusting diets based on health monitoring outcomes and new nutritional research.
- Staff training
  - Ongoing education for animal care staff on species-specific nutritional needs and feeding techniques.
  - Training on food safety and storage practices.

- Record-keeping
  - Detailed logs of daily food intake, behavioural observations during feeding, and any dietary adjustments.
  - System for tracking food inventory and expiration dates.
- Review and update
  - Schedule for periodic review of the nutrition program to incorporate new scientific findings and improvements in animal nutrition.

Foundational to this nutrition program are the written nutritional requirements and the written feeding strategies. These two aspects define not just the animal's nutrition, but also their daily schedule, their enrichment program, and their activity level. Let's look at each:

- *Nutritional requirements and calorie budgets*

  Key to any good nutrition plan is a calorie budget. Calorie budgets ensure that each individual is consuming their required amount of energy per day. At the same time, it ensures that individuals are not being over fed. Calorie budgets can be complex because they are species-specific, group specific, and individually visually specific. Because of this, the following factors are utilised to devise a calorie budget: species-typical intake; degree of activity of a population (largely determined by the constraints of the artificial environment); and the individual's body condition score.

  Understanding the species-typical intake of an animal gives a care technician a very general idea of how many calories per day an individual should consume. For some species, a care technician may have access to a species care manual – for example, an accrediting body such as EAZA may have a standard care manual for species that are typically housed in zoos. However, there are many species where no such manual exists. For these, the care technician must rely on the published data regarding that species and their ecology. This will give the care technician an idea of how much a species consumes and when they consume it. For example, lemurs may eat more during crepuscular hours (dawn and dusk). Some animals may go through natural periods of torpor where metabolic rate slows.

  The constraints that an artificial environment can have on a population's movement and activity level will profoundly determine the calorie budget that an animal requires. For example, a bird living in an aviary, where flight is limited, is going to expend less calories than a bird in the wild who can fly exponentially further. Obviously, this will create stark differences between the calorie budget of a wild bird versus a bird living in an aviary.

  At the same time, a well-crafted artificial environment can have avenues for energy expenditure to compensate for what is lacking when compared with a natural habitat. For example, chimpanzees who in the wild will typically range many kilometres per day, may have climbing structures and enrichment programs that elicit a similar energy expenditure as wild chimpanzees.

  In some cases, the artificial environment may create a lifestyle where even more energy is used than in the wild. This can be the case in enclosures such as aquariums, where even the strength of a current can be different from what is found in the wild.

All of these population variables must be taken into account when crafting a calorie budget. Oftentimes, when a new environment is created, care technicians must closely monitor the BCS of their resident populations to ensure that the calorie budget they've created matches the population's needs.

Finally, individual variables round out how a calories budget is created. These variables include age, weight, activity level, and health status. For instance, younger animals or those with higher activity levels typically require more calories to support growth and energy expenditure. Conversely, older or less active animals may need fewer calories to prevent obesity. Health status is another critical factor; animals with certain medical conditions may require a specialised diet with a specific calorie intake to manage their health effectively. Additionally, reproductive status influences caloric needs, with pregnant or lactating animals requiring a significant increase in calories to support foetal development and milk production.

Perhaps most importantly, an individual's current BCS is used to craft their calorie budget. Veterinarians may prescribe a lower calorie budget for overweight individuals, while underweight individuals may have higher calorie budgets (and prescribed dietary items). By carefully considering these individual variables, nutritionists and veterinarians can tailor a calorie budget that optimises each animal's health and well-being in managed care.

o Manufactured staple diets: For most animals in managed care, there exists manufactured stapled diets that are specifically designed to meet the nutritional requirements of that species. They've usually been designed by expert nutritionists and, indeed, they do a good job at ensuring that animals receive everything they need for regular health. However, there is a problem. These processed manufactured diets are usually fairly boring to eat. This is magnified when this is *all* that is provided to an animal. Though some have been designed to have a taste that mirrors preferred foods, they rarely fully match the level of preference that a natural food would elicit in an individual.

  Though highly effective from just a veterinary perspective, a diet consisting solely of manufactured staple foods is the antithesis of an enriching and thoughtfully prepared diet. There is, however, a happy medium. Care technicians can utilise these diets as a *portion* of the animal's diets – supplementing a base of the manufactured staple diet with actual preferred foods that an animal would eat in the wild.

  For example, a frugivorous primate may get a diet that consists of 30 per cent of staple diet and 70 per cent of fresh produce that closely mirrors a wild diet. All of this can be fed in enriching ways that allow the primate to exhibit the natural behaviours they're prone to exhibit. At the same time, due to the regular staple diet, a care technician can be assured that the primate is getting a balanced and nutritious diet.

o Intake sheets: A veterinarian may prescribe an **intake sheet**, or daily record of all food items an animal consumes along with the calorie count of each item. This may be prescribed due to animal health concerns, low BCS scores, or any number of other reasons where it would be helpful to know exactly how many calories an individual is consuming each day. Apart from these prescribed events, it is helpful to occasionally keep intake sheets on even individuals in good health.

Maintaining intake sheets for individual animals is important for several reasons. Firstly, these sheets provide a detailed record of an animal's dietary intake, including the type, quantity, and frequency of food consumed. This information is crucial for monitoring the animal's nutritional status, ensuring they receive a balanced diet tailored to their specific needs. Secondly, intake sheets help identify changes in eating habits, which can be early indicators of health issues. A decrease in appetite, for instance, may signal illness or stress, prompting timely veterinary intervention. Additionally, these records support the customisation of diets for animals with special requirements, such as those with medical conditions requiring specific nutritional management. Furthermore, intake sheets facilitate communication among the care team, ensuring consistency in feeding practices and enabling informed decisions about dietary adjustments.

- *Feeding strategies*

The manner in which an animal is fed and the time (or times) of day that an animal is fed has both direct and indirect implications of almost every other aspect of that individual's life in managed care. Feeding strategies include how often an animal is fed, what time of day an animal is fed, and the manner in which an animal is provided food. As with everything else, species-specific variables, population-specific variables, and individually specific variables must all be taken into account when determining feeding strategies.

It is important to feed animals in a manner that closely replicates how an animal would obtain food in the wild. Most of the hallmark traits of a species have developed due to the novel way they acquire resources. Animals that are merely handed food are done a disservice in that they are not able to fully realise these hallmark traits.

In managed care, feeding animals involves a variety of methods tailored to mimic natural behaviours and meet nutritional needs. Scatter feeding, for instance, distributes food across an enclosure, encouraging animals to forage and explore, thus promoting physical activity and mental stimulation. Puzzle feeders or food puzzles present another method, requiring animals to solve a problem to access their food, which enhances cognitive skills and replicates the challenges faced in the wild. Environmental enrichment devices, such as hanging feeders or food hidden within objects, stimulate natural hunting and foraging behaviours, providing both a physical and mental workout. Targeted feeding is used for individuals needing special diets or medications, ensuring they receive their specific nutritional or medicinal requirements without competition. Lastly, operant conditioning sessions incorporating food rewards not only aid in behavioural management and veterinary procedures but also serve as an enrichment activity, strengthening the bond between caretakers and animals.

Individual needs should also be taken into account when devising feeding strategies. Animals with physical conditions might be unable to feed the same way a population does. For these animals, special care plans should be created.

The amount of feedings per day should closely mirror how an animal eats in the wild. For example, animals that typically forage throughout their day should be provided food that can either be eaten in small increments throughout the day or provided with several feedings per day. Animals that typically have one large meal and then a long period of rest should likely be provided a larger single meal. Failure to feed animals in a way that mirrors their natural energy consumption can result in individuals not exhibiting natural behaviours, individuals becoming overweight, individuals becoming lethargic, or individuals going through periods of feeling unnaturally hungry.

Imagine alligators living in a zoo where they perform alligator feeding shows multiple times a day. In the wild, alligators typically have a large meal and go through a long period of rest. However, in such a situation, the managed alligators are provided food at multiple times in their day, resulting in individuals sometimes being extremely overweight or individuals who are extremely lethargic and sluggish.

Veterinary Care Plans

Similar to an individualised behavioural care plan, an individualised veterinary plan is critical to meeting the challenges that an animal may be facing physiologically. Examples of veterinary care plans include: supplemental dietary items, specialised environmental structures to meet mobility challenges, operant conditioning plans, and of course medication scripts.

- *Supplemental dietary items*

  A veterinarian may prescribe supplemental dietary items to address specific health needs or nutritional deficiencies observed in an animal. This could range from adding vitamins and minerals to the diet, to incorporating specialised foods (including specialised manufactured stapled diets) that support recovery from illness or surgery.

  During the course of these care plans, it is crucial for care technicians to meticulously record all food that an animal consumes, including these supplemental items. This record-keeping should detail the quantity, type, and frequency of supplements administered, alongside the animal's regular diet. Such diligent documentation serves a dual purpose: it enables the veterinary team to monitor the animal's response to the supplements and adjust dosages as necessary, and it ensures a comprehensive health record that can inform future care decisions. This practice underscores the importance of individualised care and evidence-based interventions in ensuring the well-being of animals under professional care.

- *Specialised environmental structures*

  Specialised environmental structures play a pivotal role in addressing the physiological challenges faced by individual animals. These structures, tailored to provide specific therapeutic benefits, are integrated into veterinary care plans with the aim of promoting physical well-being and facilitating recovery. For instance, structures such as climbing frames for primates or water features for aquatic animals can significantly enhance muscle strength and coordination. Care technicians are tasked with the critical role of regularly assessing the effectiveness of these environmental structures. Through meticulous observation and recording of the animals' interactions with these structures, care technicians gather invaluable data on the animals' physical and psychological responses. This ongoing evaluation ensures that the structures continue to meet the intended therapeutic goals or are adjusted as necessary to better serve the individual needs of the animals.

- *Utilising operant conditioning*

  In Chapter 9, we will explore, in depth, how operant conditioning can be used in care plans. Along with monitoring and assessing an animal's physiological welfare, operant conditioning can also be used to treat an animal's physiological challenges. Such programs can promote movement, allow for the administration of medication, and help to stop self-destructive behaviours.

- *Medications*

  The most obvious example of a veterinary care plan is a medical script. Medication is carefully prescribed by veterinarians after a diagnosis. This process is governed by precise protocols to ensure that each animal receives the appropriate medication at the correct dosage and frequency. The prescriptions and administration records should always be meticulously maintained in a medical records system, ensuring accuracy and accountability in the medication process.

  In most facilities, veterinarians enlist the help of care technicians to administer the medication to animals. Thus, communication is of utmost importance. The use of standard abbreviations is a common practice, designed to streamline communication and ensure the precise administration of care. These abbreviations, often derived from Latin, serve as a shorthand for frequently used terms and instructions. For instance, 'BID' (*bis in die*) means twice a day, and 'SID' (*semel in die*) translates to once a day, guiding the frequency of medication administration. 'TID' (*ter in die*) and 'QID' (*quater in die*) extend this to three times a day and four times a day, respectively, accommodating the needs of various treatment plans. The abbreviation 'PRN' (*pro re nata*) indicates that a medication should be given as needed, providing flexibility in response to the animal's condition. 'PO' (*per os*) specifies that the medication should be taken orally, while 'IM' (*intramuscular*) and 'IV' (*intravenous*) denote the routes of administration to be through muscle or vein, respectively. These abbreviations not only save time but also reduce the risk of miscommunication. They are universally recognised within the field, ensuring that regardless of location, veterinary professionals can understand and execute care instructions with clarity and precision.

Implications

The consequences of not assessing physiological challenges and not adequately providing for physiological needs can be immediate and severe. The consequences of wrongly assessing these challenges can make matters even worse. Because of this, it is critical that care technicians employ standard methods of evidence gathering and keen observational skills in order to determine the precise needs of each individual animal in their care.

Sometimes these methods can be non-invasive and done with little or no stress to the individual. Other times, invasive methods must be considered to give extremely precise information on the physiological situation that an individual is facing. Similarly, mitigation efforts for physiological challenges should be done in a precise and consistent manner. Not only does general welfare require some such precision, but sometimes the very survival of an individual is at stake.

References

Broom, D. M., 2001. Coping, stress and welfare. In Broom, D. M. (ed.), *Coping with challenge: welfare in animals including humans* (pp. 1–9). Berlin: Dahlem University.

Campbell-Ward, M., 2023. Quality-of-life assessments in zoo animals: not just for the aged and charismatic. *Animals*, *13*(21), p. 3394.

Mason, G. and Mendl, M., 1993. Why is there no simple way of measuring animal welfare?. *Animal Welfare*, 2(4), pp. 301–319.

Operant Conditioning

Figure 9.1 An Asian small-clawed otter engages in operant conditioning at the International Primate Protection League. (Photo credit: author).

9.1 Brandy Balls

In the hazy mist of morning, a small ship appears on the horizon. A lighthouse keeper on the shore stands up, peers through a telescope, and attempts to make out the markings of the ship. The ship bears no flag. Upon close inspection, the lighthouse keeper can tell that the ship is moving about erratically. Its mast swings back and forth. Each time the mast moves, the ship turns sharply in the other direction.

As the ship draws nearer to the shore, the lighthouse keeper can make out two figures on the deck. One is attempting to hold the mast, while the other appears to be shouting at him. Each time the mast swings the wrong way, the shouting man slaps the other man in the back of the head.

Finally, the mast swings in such a direction as to send the boat careening towards the lighthouse at a great speed. The lighthouse keeper braces for the impact of the ship. Lucking the shouting man runs to an anchor and throws it into the water. With that, the boat stops in an instant. This sends both men flying over the bow of the boat and into the water.

As the two men come up for air, the man that was shouting grabs the other man. He shouts some unintelligible worlds and then proceeds to drag him to the shore. When the two men reach the shore, they climb up. Both shiver due to the coldness of the water. The shouting man approached the lighthouse and knocks on the door.

The lighthouse keeper answers and invites the two men inside. Once inside, he offers them both hot coffee. The shouting man thanks him. The other man says nothing. The lighthouse keeper asks where the two men are from.

'Where I am from is not important', says the shouting man. 'I am on a quest to a volcano located in the far north sea. There, I will search for a plant. This plant can bring great prosperity to a group of people, and can bring me the redemption I need.'

As he speaks, the quiet man has found a jar full of brandy balls. He opens the jar, takes in the aroma, and reaches his hand in. He eats a brandy ball and smiles.

'What is this plant?', asks the lighthouse keeper.

'If I tell you, you would seek it yourself. I cannot risk that', answers the man.

'Well it is of no matter', replies the lighthouse keeper. 'I am committed to operating my lighthouse. However, you seem to have an issue with your crew mate'. With that, the lighthouse keeper points to the silent man.

The silent man, having now consumed two brandy balls, stares lifelessly into the distance. Crumbs cover his face.

'I found him stranded on an island', says the man. 'My ship had encountered stormy seas. The storm turned me around and I ended careening upon the shores of a deserted island. Well, deserted save for this man. When I found him, he was badly injured, he didn't speak much, and he had no memory of his past. I told him that I would take him back to his homeland if he joined me on my quest first. I am in need of a first mate.'

'When he first boarded my ship, it was clear that he had vast experience sailing ships. However, he was unfamiliar with my ship. However, when I have tried to teach him how to sail this ship, it is like he doesn't understand me. He hardly says a word. He hardly ever responds. I have tried to physically place his hands where they need to be. When he does something wrong, I shout at him. When he does something really wrong, I will slap him so that he understands. All this has done is made things worse. Now he seems nervous to do anything on the ship. He doesn't trust me. Oftentimes, he tries to hide from me. At this rate, I will never be able to get to the far north sea.'

At this, the lighthouse keeper smiled. 'I once had a very good teacher', he said. 'I learned from him because he was patient. Perhaps you could change your approach.'

'Please explain', beckoned the man.

'First of all', said the lighthouse keeper, 'if he isn't communicating with you verbally, perhaps you should try a different method of communication. Perhaps you can communicate by

merely indicating when he is doing something correctly. Instead of shouting at him, or hitting him when he is doing something incorrectly, perhaps you could reward him when he is successful. This will make the entire experience positive.'
'I don't even know what he likes', exclaims the man. 'I don't know what will motivate him.'
The lighthouse keeper points to the silent man, who is, once again, reaching in the jar for another brandy ball. 'He seems to like brandy balls.'

A Way of Learning

In our scenario, the frustrated ship's captain cannot communicate with his new crew mate. He desperately needs him to learn a skill, but without a common communication method, this is nearly impossible. Instead, he tries shouting at him or hitting him to try to get him to do the right thing. All this manages to do is to turn the entire learning experience into a very negative experience for both men. Because of this, the ship nearly crashes into the rocks.

Luckily, the wise lighthouse keeper has a simple solution – communicate by positive rewards. Doing so creates a shared communication system that is positive and effective. It is something that can be done with any living being that is teachable.

Imagine yourself facing an activity that you've never attempted before. This activity contains several steps (all separate actions) but has a clear end goal that would denote the successful completion of the entire endeavour. Let's imagine that you are riding a horse for the first time.

In order to successfully rise the horse you need to learn all of the steps. You need to learn how to mount the horse. You need to know how to hold the reins. You need to know how to ask the horse to move. You need to know how to tell the horse where to go. You need to know how to tell the horse how fast to go. You need to know how to tell the horse to stop. You need to know how to dismount the horse. All of these activities may be incredibly different from anything you are used to doing. The thought of doing any of this may evoke fear. Indeed, doing any of the steps incorrectly can be extremely dangerous.

The idea of climbing on to the back of a horse and allowing a horse to completely control your movements (sometimes at great speeds) is a completely unnatural thought. If you've never had experiences with horses before, it can be very frightening! Indeed, without the right guidance, if one were to just climb onto the back of a horse, a potentially deadly disaster could occur.

As with most endeavours, there are a lot of different actions that must be learnt and pieced together before you can go from being a person who has never ridden a horse before, to a person who can climb on a horse and successfully and safely get from point A to point B. In order to master these steps, you must both learn the proper way to perform these tasks, but also you must condition yourself to the process – performing the tasks with precision, overcoming fears, and seeing the endeavour as a rewarding activity, rather than a purely stressful one.

Let's now imagine that you have a good equestrian teacher at your disposal. The first thing this teacher is going to do is to teach you the proper way to ascend a saddle and effectively mount the horse. The teacher can tell you everything that you need to know in order to perform the task successfully, However, until you actually climb on to the horse and act in the instructed manner, you will not actually be conditioned to perform the task.

Let's imagine that you first try to get onto the back of the horse. You falter and stumble around. When you do this, the horse immediately begins to run. Your foot is caught in the saddle. The horse, runs at great speeds while dragging you across the ground. You end up bloody, broken, and severely injured. It is likely that the next time you attempt to get onto the back of the horse (if you even ever attempt it again), you are going to be highly more trepidatious in doing this behaviour. This reticence may cause you to make even more mistakes. Because of what happened the previous time, the consequences have now pervaded your consciousness, giving you a negative reference point to ever performing this task again. Your expectation, when mourning a horse, will be to endure the same experience you had previously.

All of this would have been different if you had been conditioned to a positive experience, rather than a negative experience. Let's imagine a different occurrence and a different outcome. Let's imagine that your equestrian teacher helps to ascend the horse properly. When you do this, the horse does not run off. Instead, you successfully get on the horse and the equestrian teacher compliments you – praising you for performing the task correctly. You have a sense of pride in your achievement. The teacher's compliments feel good. You have now been conditioned that, when you get on a horse, instead of injury and mayhem, you feel accomplishment and pride. As such, you will likely keep doing this behaviour in hopes to get such a positive reaction again.

Now that you've successfully mastered mounting the saddle of the horse, your equestrian teacher will likely give you instructions on how to hold the reins; followed by instructions on how to ask the horse to move, and how to let the horse know where you want to go. Let's again, imagine two scenarios.

In the first scenario, you begin to hold the reins the wrong way. The equestrian teacher begins to scream at you, telling you that you're inadequate, and that you'll never successfully be able to ride a horse. While this may encourage you to follow the instructions more closely next time (who wants to be yelled at and belittled in such a way?), it also may discourage you because you have now been given a negative association with performing the task. You may be so afraid of getting berated again, that you're too nervous to grab the reins.

You then attempt to give the horse a cue to move forward. However, you don't use exactly the right words or the right that the instructor gave you. The instructor again, belittles you for not following instructions. Once again, you may want to follow the instructions more closely to avoid this, but you also may want to give up riding the horse altogether because of the unpleasantness of the entire experience.

However, let's now imagine a different scenario. This time, the teacher patiently waits as you first struggle a bit to hold the reins correctly; but when you finally get the reins in your hands the right way, the instructor gives you praise and encourages you to keep going. In this scenario, the instructor has accomplished allowing you to perform the correct task and rewarding you when you perform the task correctly. You've now avoided the negative reinforcers of the first scenario. While it may have taken you more time to grasp how to hold the reins, you now have a very positive reference point onto the entire experience.

Next, the teacher gives you instructions on how to ask the horse to move and how to tell the horse to go where you want him to go. Again, the teacher patiently waits while you work on the correct cues that you are given. When you've given the correct clue, the teacher praises you and the horse follows your command. You now have been rewarded twice positively.

Soon, you have mastered the art of riding a horse. Though you may have times that are challenging, your initial experiences have been positive. You have properly conditioned yourself to the positive experience of riding a horse.

Positive and Negative Reinforcers

In both of our scenarios, you have been conditioned as to what to do to ride a horse by **reinforcers**, or conditioning stimuli that influence your actions. In the first scenario, these reinforcers are negative. You want to avoid these reinforcers. Therefore, you try to act differently to avoid them. With a negative reinforcer, you may choose to act in the 'correct' manner, to avoid the reinforcer, or you may avoid the entire activity altogether.

On the other hand, a positive reinforcer works by attaching a good stimulus onto a desired action. It may take you a bit longer to figure out what the desired action is, but once you do, you receive a positive reward. This reward makes you want to keep acting in this manner, so that you can attain the reward again.

In all of our scenarios, you have been conditioned to a defined series of actions that leads to a goal.

9.2   Defining Operant Conditioning

Operant conditioning is a complex technique in animal care, with an array of methods and techniques that could fill up volumes of books. For the purposes of this book, I intend to merely give a broad overview of the subject. In short, this is merely a primer to a much larger world of animal husbandry.[1]

You often hear the phrase 'animal training' when the process of teaching animals is being discussed. However, I feel that the term training carries with it connotations that evoke notions of *programming* animal behaviour rather than *teaching* animal behaviour. As such, I prefer to use the term **operant conditioning** because operant conditioning carries only a connotation of teaching and learning through a defined process. This is a tried and true process that has been used in animal care since the nineteenth century!

Operant conditioning is an invaluable tool in terms of the welfare of animals in managed care. The process allows animals to voluntarily participate in their own care and welfare. For example, think of the difference between an animal being darted with medicine versus an animal voluntarily presenting their body for a needle stick. Imagine an animal being conditioned to sit on a scale versus having to be anaesthetised and weighed during a veterinary procedure.

---

[1] Karen Pryor's, *Don't Shoot the Dog* is a recognised classic in this field. See Pryor, 1984.

Additionally, operant conditioning allows care technicians to gain valuable information on the animals in their care without having to do anything invasive, rather by participating in an operation that can usually be very enriching for the animal. Think of an animal that is conditioned to sit still for an ultrasound. Imagine an animal that has been conditioned to open his mouth on command, so that the care technician can examine his teeth.

Operant conditioning can also greatly aid in basic husbandry practices. For example, animals can be taught to come into their night houses at a given command. Animals can be taught to station at a certain area during feedings. Animals can be conditioned to nest in certain areas of their night houses.

At its core, operant conditioning simply means to modify a behaviour by reinforcement. This reinforcement can be either positive or negative. Both work to condition the learner through the association between a behaviour and its consequence. Operant conditioning focuses on how an action leads to an outcome. By associating specific behaviours with positive or negative consequences, the learner can be conditioned to repeat or avoid certain actions.

Operant conditioning works in the following manner. There is a behavioural goal that a care technician would like to see an animal achieve. This behavioural goal may be to increase a desired behaviour. It may be to decrease undesired behaviours. If it is to increase a behaviour, the care technician chooses whether or not to reinforce this behaviour by a positive stimulus that follows the desired behaviour (positive reinforcement) or by adding a negative stimulus that remains until the desired behaviour is achieved (negative reinforcement). In the case of decreasing an unwanted behaviour, the care technician may choose **punishment** if the behaviour is exhibited, or the process of adding negative stimuli, or removing a positive stimulus, if the undesired behaviour is exhibited.

Contrasting Reinforcement with Punishment

There is a major difference between reinforcing behaviours using punishment to achieve behavioural goals. Reinforcement increases desired behaviours, while punishment aims to decrease or extinguish unwanted behaviours. In most cases, reinforcing desired behaviours, and conversely replacing undesired behaviours, is a far more effective approach than punishing unwanted behaviours.

As we have explored, positive reinforcement introduces a pleasant stimulus following a behaviour, enhancing the likelihood of its recurrence. For instance, rewarding your dog with a treat for sitting on command exemplifies positive reinforcement. In most cases, positive reinforcement is a pleasant experience for the individual being conditioned.

Negative reinforcement, on the other hand, can be an unpleasant experience. It is often confused with punishment. However, whereas punishment aims to decrease or extinguish a behaviour, negative reinforcement, like positive reinforcement, aims to increase a desired behaviour. It works by removing something unpleasant when the desired behaviour is achieved. For example, a loud sound may be turned off when a task is completed.

Punishment, like negative reinforcement, is an unpleasant experience for the individual being conditioned. Punishment, like reinforcement, can come in two forms – positive and negative. Positive punishment adds an unpleasant stimulus to reduce a behaviour's occurrence. For example, a care technician turning a water hose on a chimpanzee to stop aggression is positive punishment. Negative punishment, on the other hand, entails removing a desirable stimulus. For example, taking a bone away from a dog for misbehaviour, is a negative punishment.

From a welfare standpoint, positive reinforcement is by far the most sound method of operant conditioning. While negative reinforcement and punishment can be effective tools in conditioning, they can lead to unnecessary stress, heightened anxiety, and can create distrust between an animal and a care technician. Therefore, utilising positive reinforcement should always be a priority.

Positive Reinforcement Training

I once had a colleague who worked as a zoo registrar. She absolutely detested her job. She once described it to me as tedious, lifeless, and without reward. She claimed that she couldn't sleep at night because the sound of a ticking alarm clock and creaking house was more stimulating than her daylight hours of staring at animal records. Indeed, she had the unenviable task of having to digitise individual animal records, some of which dated back 30 years. Day after day, she came into her office, sat down at her desk, and trudged through the endless files of animal records. Her only reward was a distant paycheque. However, the paycheque was far from immediate and thus a fairly intangible reward while knee deep in veterinary files and identification numbers.

After years of this mundane work existence, she realised that either something had to change or she needed to find a new career. It was at this point that she developed an innovative approach to her own job satisfaction. The gift shop at the zoo sold bags of cheese-flavoured popcorn. The popcorn was highly processed, artificially flavoured, very unhealthy, but *really* tasty. Having devoured several bags of it myself, I can attest.

Each morning, the zoo registrar would go to the gift shop, purchase one of these bags of savoury salt and fat, and bring it back to her office. She would then discipline herself to the following rule: each time she finished a section of an animal's records, she would allow herself one handful of the cheese popcorn. At no time could she allow herself to eat any popcorn, unless she had completed a section.

When she was hungry, or particularly craving more popcorn, she would work even faster to complete the tasks. She was even able to track how much work she had accomplished by how much popcorn she had consumed.

By rewarding herself in this way, the quest to finish these animal records became quite a bit of fun. Her productivity increased, as did her job satisfaction. In essence, what this registrar had done was to condition herself to the behaviour required in her job through a means of positive reinforcement. Not only did her job performance improve but also her general well-being and satisfaction with how she spent eight hours of every day.

What is important to understand is that positive reinforcement conditions the experience of performing a task. The satisfying moment is associated with not just receiving the reward but also the steps it takes to achieve the reward, thus conditioning the entire experience into a positive one. This can occur when one is learning something new or performing something that has been done time and time again.

Positive reinforcement can make a mundane task, such as the registrar's tedious job, or even a scary or painful experience, such as getting a needle stick, something that is associated with a gratifying reward. It can, in fact, be so effective at overshadowing what would normally be a negative moment, that the entire experience is positive and satisfying – even something that is happily anticipated.

Because positive reinforcement training (PRT) works by incentivising desired behaviours by providing a favoured reward, rather creating and removing a negative stimulus when the desired behaviour is achieved (negative reinforcement), or inflicting a negative stimulus when an undesired behaviour is performed (punishment), it ensures a gratifying experience that an individual generally wants to do again. Thus, PRT is the most effective method of operant conditioning.

In animal care, PRT is both a husbandry tool and a welfare tool. Allowing animals to voluntarily engage, *in a positive and rewarding* manner, in such tasks as being weighed on a scale, or receiving an injection, can remove the undue stress, while allowing care technicians to perform critical husbandry tasks. Additionally, animals typically look forward to such sessions and can be enriched by the entire experience.

9.3   PRT Techniques

In simplest terms, PRT works by creating rewardable steps towards a larger goal. Each time a step is performed correctly, an animal is rewarded. For example, imagine if you are attempting to train a chimpanzee to receive a hand injection. The first thing you would need to do is to have the chimpanzee sit in an area where it is possible to reach him. This would require having the chimpanzee to approach a specified area and sit. When the chimpanzee moves towards this area, he is rewarded. When he sits, he is rewarded. When he remains still, he is rewarded. This behaviour is repeated again and again, until the chimpanzee understands the command, and reliably responds correctly. Once he is proficient in sitting and remaining in the desired area, he needs to hold his arm up to the mesh and remain still. Again, there are several steps to this. Each time he turns his body in the correct fashion, he is rewarded. Each time he presses his arm to the mesh, he is rewarded. Each time he holds his arm still for a prescribed period of time, he is rewarded. Finally, he needs to be able to receive a needle stick. When he allows a needle-less syringe to be pressed to his arm, he is rewarded. When he allows a syringe with a needle to be placed close to his arm, he is rewarded. When he accepts a needle stick, he is rewarded. When he allows medicine to be injected, he is rewarded.[2]

So in this scenario, the behavioural goal is to have a chimpanzee voluntarily accept a needle stick. While the chimpanzee is learning to do this, the care technician

---

[2] See Schapiro et al., 2003.

is **shaping** the behaviour, or conditioning the steps necessary in order to reach the larger behavioural goal. Once the behaviour is shaped, it needs to be routinely practised. For behaviours that have already been shaped, the care technician must work on **maintaining** the behaviour or routinely practising a conditioned behaviour that has already been shaped.

Within this technique, there are three critical rules to follow:

1. Consistency: Commands must be consistent. Behaviours that garner rewards must be consistent. As with most everything in this book, consistency can be greatly aided by having a written plan.
2. Quick delivery of rewards: When an animal performs the desired behaviour, rewards must be delivered quickly and clearly as a result of the desired behaviour.
3. Participation must always be voluntary: When an animal loses interest in the session, the session must end. Attempting to coax an animal to participate or showing frustration when an animal loses interest, can work against the overall goal. Also, openly showing frustration can actually be a form of punishment, thus undermining the entire method of positive reinforcement.

Commands

A command is a consistent signal that is used to elicit a desired behaviour. Command can be audial (such as verbal or a specific sound), visual (such as a gesture or the flash of a light), or tactile (such as a touch). The command is a very specific directive – meaning, that it must always signify the desired behaviour and nothing else.

Introducing a command commences in the following manner:

- Introduction of command: The trainer introduces the verbal command. For example, if you were training a chimpanzee to grab a bar, the command might be the verbal command 'bar'. The command is given while using a target stick to guide the chimpanzee's arm up to the bar. The target stick also will serve as a visual cue – which aligns with the verbal command and clarifies the desired behaviour.
- Immediate positive reinforcement: As soon as the chimpanzee lifts his arm towards the bar, in response to the command, the chimpanzee is rewarded. This immediate reinforcement helps the chimpanzee associate the command with the action and the reward.
- Gradual removal of physical guidance: Over time, the care technician reduces the reliance on the target stick, focusing more on the command.
- Consistent practice and reinforcement: The care technician consistently practises this command, always providing positive reinforcement for the correct response.

Rewards

At its core, a reward in PRT is anything that an animal finds pleasing and is willing to work for. These rewards can be categorised into several types, including food rewards, social rewards, and environmental rewards.

Food rewards are perhaps the most straightforward and universally understood; these can range from a small piece of a favourite treat to a full meal, depending on the training session's requirements and the animal's dietary needs.

Social rewards involve positive interactions with the trainer or other animals, such as petting, verbal praise, or even playtime, which can significantly reinforce desired behaviours by fostering a strong bond between the animal and the care technician.

Environmental rewards offer a change in the animal's surroundings as a form of reinforcement. This could include access to a favourite resting spot, the opportunity to explore a new area, or the introduction of new toys or enrichment items. The effectiveness of environmental rewards lies in their ability to satisfy the animal's natural curiosity and desire for stimulation.

In PRT, the key to success is, like so many other things in animal care, the ability to tailor to the individual. Individual preferences will determine the most effective category of reward, and specific reward.

Bridges

A **training bridge** is a signal that is used to communicate to the animal that the specific behaviour is correct and will be rewarded. For example, a care technician may sound a clicker or whistle, in conjunction with a reward, each time a desired behaviour is given by the animal. This signal allows the animal to understand that this behaviour is the desired behaviour. A training bridge can be delivered more quickly than an actual reward. In time, the training bridge can be substituted for the actual reward, allowing care technicians to actually produce the rewards less frequently as the behaviour becomes more commonplace.

9.4   Examples

Station Training

Station training is often the precursor to more elaborate behaviours such as scale training or recall training. It simply means to train an animal to station themselves at a predetermined location. Often, implicit in this training, is for the animal to remain at this location for a given period of time. When the animal does this, a reward is given.

Target Training

Like station training, target training is usually a precursor to more elaborate trained behaviours such as hand injection training or body part presentation. Target training an animal means to teach an animal to present one of their extremities to a predetermined point. Training is often done with an object (such as a plastic ball attached to a stick). A command is given to touch the target with an extremity. When this is accomplished, the reward is given.

Recall Training

Recall training occurs when animals are trained to return to an enclosure. For example, animals may get recall training at a zoo, where they are on an outdoor exhibit during the day, and need to return to night housing at the end of the day. Recall training can also be utilised during escape responses.

Injection Training

Injection training is often a combination of several behaviours such as target training and station training. In injection training, an animal learns to voluntarily present a body part and receive a needle stick. There are often many steps in training this behaviour. As such, injection training can be one of the more complicated and difficult behaviours to shape and maintain.

Scale Training

Scale training occurs when an animal is trained to either sit on a scale or hang from a scale in order to obtain a reliable weight. This behaviour can sometimes be fraught with difficulty – for example, some animals, particularly primates, are prone to holding on to a wall or mesh as they stand on a scale, and thus giving an inaccurate weight. Therefore extinguishing these behaviours is critical for successful scale training.

Implications

Operant conditioning, or the conditioning of voluntary behaviours, is the foundational maintaining a care goal of allowing opportunities for freedom of choice for animals in managed care, while providing for critical husbandry and veterinary needs. It provides for this goal by allowing animals to voluntarily participate in their own critical care. In turn, this reduces stress and provides a more effective way for care technicians to take care of animals.

The most effective method of operant conditioning is PRT. PRT conditions voluntary behaviours by providing positive reinforcers when the desired behaviours are achieved (rewards). This stimulus encourages behaviours that the car technician would like to shape and maintain.

A robust PRT program involves staff shaping and maintaining behaviours as part of their daily husbandry routine. If done properly, PRT can be done with very little additional time to a daily routine.

The benefits of a robust PRT program are as follows:

- **A boost to general animal welfare** with regard to a reduction of stress and anxiety as animals voluntarily participate in their own care.
- **An ease of veterinary operations.** Animals can be trained to receive medication, anaesthesia, and station for visual examination.

- **A reduction of sedations.** A veterinarian can examine an animal in a more non-invasive manner.
- **The ability to move animals with less stress.** Animals can be trained to station at certain points to allow them to shift into desired places with less stress. This can also help in the case of an escape.
- **A highly enriching activity for all animals.** Animals generally enjoy participating in PRT exercises.
- **A highly rewarding activity for the staff.** Participating in PRT can help staff satisfaction, as they learn new skills that are rewarding to execute.

Case Study: Using Positive Reinforcement Training to Improve Heart Health for Great Apes in Accredited Zoos and Beyond

Tina Cloutier Barbour, PhD, Associate Vice President of Animal Care and Welfare, Dallas Zoo

On Her Career Path

I have always loved animals. This passion manifested itself when I was just two years old, and my parents put me on a horse for the first time. After that, I was absolutely hooked. So it was natural for me to pursue a career path that would allow me to contribute to animal welfare in a meaningful, positive way.

Figure 9.2 Dr. Cloutier Barbour surveying the chimpanzee yard at the Dallas Zoo. (Photo credit: author).

I have also always been fascinated by people – largely because I find them more difficult to understand than animals. This dual interest led me to pursue a degree in Anthropology, where I eventually found myself volunteering at a local zoo. My role there was to collect data for a PhD project that was investigating chimpanzee vocal communication. It was a year of solitary, tedious work, which left plenty of time for reflection and observation. Those countless hours spent quietly observing a large, multi-generational group of chimpanzees – getting to know their individual roles and personalities – were transformative. That experience ultimately shaped the rest of my life. I decided to pursue a career in primatology, and promptly got a job as a zoo-keeper, which I worked alongside earning my Master's degree. The overwhelming feeling of awe that I felt the first time I met those chimpanzees up close is something I will never forget.

One was especially fascinating: Little Mama, who was, at the time, the oldest living chimpanzee in human care. To know Little Mama was to love her. I had spent a lot of time watching her, and her interactions with the rest of her large social group. It wasn't long before I noticed that she didn't experience the trademark swelling that female chimpanzees display to signal ovulation. Given her age, and the fact that chimpanzees are our closest living relatives, I immediately wondered if chimpanzees experienced menopause. No one seemed to know, and I was so curious that I eventually found myself at the University of Utah, where I earned my PhD in Evolutionary Ecology. My research centred on the evolution of human life history, using chimpanzees as a comparative model. I became an expert in chimpanzee aging, which nicely supplemented my behaviour-based understanding. My time in university was fascinating and satisfying, but I desperately missed working with chimpanzees. When the chance to go back to that zoo – and those chimpanzees – presented itself in the form of Primate Curator for the area, I jumped at it.

On Her Role as Primate Curator

Stepping out of the academic realm and back into the world of animal management was humbling. Here, I was charged with the care and welfare of 19 chimpanzees and 2 white-handed gibbons, among other primates at the zoo. As Curator, my decisions must centre around intimate knowledge of the animals in my care, while also keeping a bird's eye view of the program under consideration. This meant simultaneously managing the daily tasks that accompany the care of multiple groups of incredibly complex apes – diet evaluations and changes, health status reviews, geriatric care, social pressures and their behavioural impacts, and so on – all while operating within the context of evolving team dynamics, conflict resolution, program sustainability, keeper safety, and so on.

It was a responsibility that I took very seriously; I recognized that a single mistake or oversight could mean serious injury or harm to an animal that I cared for deeply or a member of my team. This also meant that I was constantly seeking out opportunities to improve the welfare of the animals in our care. Luckily, graduate school had left me well-suited to the challenge. By the time I graduated, I was formally trained in project management, as well as large-scale, big-picture thinking.

Figure 9.3 At the entrance of the chimpanzee forest at the Dallas Zoo. (Photo credit: author).

On Positive Reinforcement Training

One of the best, most effective tools in a 'positive welfare' toolbox is PRT. PRT is defined as the 'presentation of a reinforcer following performance of a correct response which increases the probability that a response will occur in the future' (Zeligs, 2014). Essentially, it is a learning strategy that rewards targeted (i.e., desired) behaviours with something considered valuable to the animal. A chief tenant of any good PRT program is choice and control. This simply means that if an animal does not want to participate in a training session, they are given the choice to walk away without any negative consequences, or punishment. While PRT has shown to benefit virtually all animals, this is *especially* true of great apes, who are not only incredibly intelligent but also have fantastic memories and strong opinions about fairness.

Positive reinforcement training is a very appreciated and often-used tool for medical or husbandry behaviours for zoo-housed animals. For instance, we often use PRT to train animals to voluntarily accept an injection or allow a blood draw. In species or taxa with well-known health conditions – such as great apes, which struggle with cardiovascular disease – we might prioritize voluntary hearth health screening. All of this minimises the need for invasive medical procedures, which in turn significantly reduces stress on all participants … animals and humans alike.

As the Primate Curator, I was always looking for ways to improve the welfare of the animals in my charge. I was also always searching for ways to reduce stress in the keeper's lives. The development of a proactive, innovative PRT program featured heavily in my bid to achieve both goals. And so, when I happened to see an infomercial late one night for a new FDA-approved device that could detect arrhythmia and arial fibrillation in humans, my first thought was, 'I could train a chimp to use that.'

The next morning, I reached out to the Great Ape Heart Project to determine whether this had been done before. Would application on chimpanzees even be

Figure 9.4 At the Tonkolili Chimpanzee Site, a conservation project Dr. Cloutier Barbour co-founded with the author. (Photo credit: author).

functional? It made sense to me, as we are all great apes, but I was interested to hear from the heart health professionals. The answer that I received was exhilarating: no, they hadn't heard of anyone else trying it, but there was no reason to expect that it wouldn't work on a chimpanzee. And: they were very interested to hear how it went if I pursued this line of investigation.

I immediately requisitioned the product and started developing a shaping plan. A shaping plan is a document that trainers use as a roadmap to success. We have to carefully think about each step, or approximation, in the training process: what antecedent behaviours the animal needs to know before proceeding to the next step in the process, what conditions the behaviour will occur in, what early refusal looks like in that species, what tools we will need along the way, and so on. Each step to get us to the final behaviour is carefully documented and planned for. And then we get to work.

For me, training has always been one of the absolute joys of my career. I will never forget the feeling of delight and accomplishment that resulted from my first successful training session with a chimpanzee. As a young zookeeper, I was working with our alpha male, Higgy, on establishing a 'trade' behaviour. After a few short sessions, he understood the ask and handed me back an object that I had placed in his possession. As soon as he handed it back to me when cued, I was so overcome with excitement that I immediately shouted, 'Good!' Higgy leapt up and started laughing (yes, chimpanzees laugh). I can't say for certain what he was feeling in that moment, but I was so proud of us both. It was another transformative moment in my life – training was forever joyful for me. So it made sense that I would ask Higgy to once again be my partner in this new behaviour so many years later. He was often my 'go to' partner when I was implementing a new training plan. Training often requires trust, especially in chimpanzees, and we had built years of it over these sessions.

And so, we dug in. As always, Higgy was a fast learner – always engaged and eager. The training progressed rapidly, and we were able to get a successful reading on him within a very short timeframe. Any minor bumps in the road were addressed along the way, and we were quickly prepared to start with our next chimpanzee. Meanwhile, those results were reported back to the Great Ape Heart Project, as well as our veterinarians. It had worked! The device was able to successfully contribute to non-invasive heart health monitoring in chimpanzees.

I shared the training plan and device information among the community widely, as it was another useful tool in the toolbox to use to fight cardiac disease in the species (Cloutier Barbour et al. 2020). To date, and to the best of my knowledge, it has been successfully applied across all great ape taxa, and is used among facilities worldwide as a heart health monitoring tool. Zoos, sanctuaries, and laboratories alike have added it as a priority behaviour in their PRT programs due to its accessibility and health surveillance benefits.

It was a bit of a long and winding road that got me where I am today, but I wouldn't change any of it. I consider it an absolute privilege to be able to positively impact the care and welfare of animals and care staff daily. Sometimes, it's a difficult road. Meaningful work with animals is *hard*, both physically and emotionally. It can feel like a thankless job some (or most) days. But the days that you can actively partner with an animal to improve their welfare ... when you find yourself shouting, 'Good!' with childlike joy, and a *chimpanzee* leaps up and laughs with you? When you know that your work has made a positive impact on the animals and humans around you? Those days are hard to beat. Those days will forever transform you and the way you look at the world. Animals have shaped my life in so many ways. Animals have helped me thrive. I am just grateful to return the favour any chance I get.

References

Cloutier Barbour, C., Danforth, M. D., Murphy, H., Sleeper, M. and Kutinsky, I., 2020. Monitoring great ape heart health through innovative electrocardiogram technology: Training methodologies and welfare implications. *Zoo Biology*, *39*(6), pp. 443–447.

Pryor, K., 1984. *Don't shoot the dog!: How to improve yourself and others through behavioral training*. New York: Simon and Schuster.

Schapiro, S. J., Bloomsmith, M. A. and Laule, G. E., 2003. Positive reinforcement training as a technique to alter nonhuman primate behavior: quantitative assessments of effectiveness. *Journal of Applied Animal Welfare Science*, 6(3), pp. 175–187.

Zeligs, J., 2014. *Animal Training 101: The complete and practical guide to the art and science of behavior modification*. Minneapolis: Mill City Press.

Part IV

Operations

Curatio Fundamentorum **6:** There is an inherent risk in caring for another species. A culture of safety protects both the caretaker as well as the cared for.

10 The Culture of an Animal Care Organisation

Figure 10.1 A sea turtle in a mixed species exhibit at the South Caroline Aquarium. (Photo credit: author).

10.1   A Letter from the Sea

Dear Vicar,

I am writing to you to update you on my progress. It has been months since we spoke in the tavern. I appreciate the blessings before my quest and I wanted to return the favour by keeping you apprised of my journey.

As you know, I left on a quest to find a mysterious island where a special juniper tree grows. Once I find this tree, I can sell the products of its berries for gold. Once I have this gold, I can bring wealth to the tiny village I ruined. This quest has taken me to the far north seas.

During my travels, I have encountered great peril. The stormy seas have set me adrift and off course. At one point, I was careened on a desert island. There, I met a stranded castaway. Though this man did not talk, I was able to convince him to come aboard. Luckily this man is an experienced seaman. As such, I have successfully been able to teach him how to operate my ship. After some time, he began to speak. He tells me that he was once on a ship called The Black Valentine. The ship wrecked upon the island and he has been there ever since. He understands my quest and has agreed to help me. In return, once this quest is over, I will return him to where he once lived.

It was several weeks at sea before we encountered our first major issue. My ship's wheel broke – leaving us completely adrift. We disassembled it and found that the tiller ropes had snapped the pedestal and destroyed the spindle. Luckily, when the wheel broke, we could see the lights of a nearby port on the horizon. We were able to successfully rudder the ship to this port.

When we arrived, we found a depressed port village. The community was very unwelcoming. They met us with extreme suspicion. We soon found out why. Their beloved king had taken ill. He was at the brink of death when we arrived. No one in the town knew how to cure him.

We told them that we needn't stay in town long. Our ship was damaged and we just needed some simple repairs. We asked if anyone in the town was a boatswain and knew how to repair a ship's wheel. They sneered that no one in the port could fix our ship. Finally, we agreed to pay them if they could take us to someone. One of the men in the village said that he knew of one boatswain who lived far up the coast. He could bring us to him, but it would cost us 50 gold pieces for him to take us. When we arrived the boatswain would charge us 50 additional gold pieces for his labour. This was 100 more gold pieces than we had.

My companion and I retreated back to our broken vessel. As we pondered our situation, I came up with a dishonest scheme. I found an old bottle of gin. I squirted some squid ink into it, added some finely chopped tobacco pieces, and shook up the bottle. The smell of this concoction was terrible. However, it did smell like some sort of medication. I grabbed the bottle and told my companion to follow me.

When I disembarked the boat, I found a crowd of townspeople. I asked for their attention and exclaimed that I had an elixir that could cure any disease. If they agreed to take us to the boatswain and pay him to repair our ship, I would give them the elixir. Then they could save their king – but only after our ship was fixed.

At this, they shouted at me and threatened to kill me if I didn't give them this medicine. Two men tackled me to the ground. They held me down and forced me to produce the elixir. I told them it was in my pocket. They snatched it from my pocket and dragged me and my companion to the palace of the king. They told me that they would give him the elixir. If he recovered, they would see to it that our ship was fixed. If he did not, they would hang both of us.

Before I knew it, I was standing at the foot of the king's bed. The king's servants gave him my dishonest cocktail of squid ink and spirits. The king gagged and spit it out. They tried again. The king swallowed the portion. He gasped and made a choking sound.

I looked at my companion. We would soon be facing the gallows.

However, something miraculous happened. The king sat up. Colour returned to his face. He told his servants that he felt better!

The king asked who we were. We told him of our plight. The king called for the boatswain to come to us. He told us that they would repair our ship and make it better than it had ever been before.

The town celebrated. My companion and I were treated as heroes. That night, they held a banquet in our honour. They provided us lodging in the palace as we awaited our boat repairs.

As the days went by, I began to notice things about the town that I thought could be improved. Things in the town could be done more safely. Things in the town could be done more efficiently. Things in the town could be done more how I had been taught to do things. I began to think that maybe I could use this time to truly help these people. Perhaps this was my path to redemption.

Perhaps this was my quest all along.

I began by going to the shipyard. I showed the workers there how to tie their sails to the mast better. I told them how to properly swab the decks.

The next day, I went into the kitchens. I told them better ways to prepare their foods. I told them better ingredients to use.

I went to the fisheries. I looked at what the fishermen were catching. I showed them that there were bigger and more colourful fish they could catch and eat.

Because I had saved their king, everyone listened to me. I was able to create a wonderful new world for this depressed port town. The townspeople looked to me as if I, in fact, were their king. The sails were tied to the mast in the way I had shown them. The food was prepared with the ingredients I had chosen. The fish that were caught were the large colourful fish I had found.

On the morning that the repairs on my ship had been completed, I decided that I would remain in this town. I decided that this was the place of my true calling. I decided that the entire quest must have been God's way of directing me here. I would

*continue to create a world where these people could live better lives. But little did I
know that terrible things were about to occur.*

*First, a report from the sea came into the port. One of the ships had lost their sails.
They had become untied. The men on the ship were swept away by a squall they could
not avoid. There were no survivors.*

*Next, several of the people in the town began to suffer severe stomach pains. The
food they had been given, prepared in the way I had recommended with the ingredients that I had chosen, had given them an allergic reaction. The hospital soon filled
with extremely ill townspeople – many of them children.*

*Then, it was discovered that the large colourful fish was actually toxic. This was
discovered because the first man to eat one collapsed in agony. Within minutes, he
was dead.*

*But the worst was about to occur. As I stood in the center of town, devastated by
what had befallen these people based on my doing, I heard the sound of a horn. The
sound was coming from the king's palace. The sound was followed by screams.*

The king was dead.

*The king's illness had not been cured. His brief improvement was nothing more
than a coincidence. Nothing in the squid ink, tobacco, and gin had helped the king.
He had still been sick. He had still been dying. Now he was gone.*

*At the news of the king's death, I looked around. The townspeople had gathered
in front of me; their eyes glistened with hatred. I had not done anything to help these
people. I had not created a world where they could live better lives. I had lied to them.
I had conned them. Now, they wanted me dead.*

*I yelled to my companion. We ran to our ship. The townspeople chased after
us. Luckily, we outran them. We boarded our vessel and shoved off, out to sea. We
could hear the townspeople screaming at us from the shore. We lowered our sails
and sped away.*

We are, once again, headed north.

Forgive me, Vicar. I am nothing more than a scoundrel.

With kindest regards,

-The Mariner

What Is Culture?

While the mariner thought he was doing something wonderful for the townspeople,
he had no actual understanding of their world. He had no understanding of the king's
actual illness. He had no understanding of why the townspeople traditionally ate they
food that they did. He had no understanding of why the sailors employed a certain
technique in tying their sails to the masts. There are environmental factors, biological factors, and social factors that he didn't contend with. In the end, nothing he did
worked – and he was found to be the fraud he was.

The Mariner's failure is steeped in the fact that he had no understanding of what
culture is. Populations have certain beliefs, practices, and skills based on what has been
passed down from generation to generation. These aspects, or traits, exist because they

worked in the past, and continue to work in the present. This is a population's culture. All populations have culture. There are national cultures, local cultures, and yes, there are cultures within organisations and facilities – including animal care facilities.

Before we can even begin to discuss this concept, we must define it. An anthropologist might define **culture** as what we learn from our population, rather than what we genetically inherit. When it becomes passed down and 'fixed' within a population, it is culture. A culture exists in every human population. It defines how we respond to any given stimuli by what we've learnt from each other. Cultural traits include our skills, our concepts of right and wrong, our perceptions of what we find beautiful, our perceptions of what we find vile, our concepts of how the world operates, and even our motivations for our daily routines. Culture helps define how you acquire resources, how you stay safe, how you coexist with others, and how you see the world. In short, culture is what you learn from your population that allows you to survive. Culture exists everywhere there are humans – *again, this includes animal care facilities.* Like everywhere else, each facility has certain practices, skills, beliefs, etc. that have been passed down from one generation of staff to the next.

All human populations function through culture. There is a culture within a geographic population, a culture within a social group, and a culture within an organisation. The culture at an animal care facility is a collective of the learnt behaviour within the organisation that has become fixed over time. Beyond just the methods and protocols for the workplace, this can include general attitudes towards everything from how the job is approached to how staff treat each other and animals. This can include a general acceptance of the mission of the organisation. This can include a general feeling of trust (or lack thereof) of the leadership within an organisation. Perhaps, most importantly, this can include a general approach to safety.

Cultural traits are the various facets of a culture that make up the collective culture. Each cultural trait is a cog in a wheel that can largely determine the success or failure of an animal care program. Cultural traits at an animal care facility appear and evolve like any other cultural trait. They may appear when a need is encountered. For example, someone may get hurt on the job. A facility may review what caused the injury and implement a protocol that helps to prevent the injury in the future. Over time, as the staff get more familiar with the protocol, they may make small improvements. Eventually, the protocol may look very different from how it appeared at its onset. This is an example of a positive appearance and evolution of a cultural trait. It appeared through a measured response to a need. It evolved due to experience and proficiency. However, not all cultural traits appear and evolve in such a positive fashion.

Cultural traits can also occur that are *maladaptive* to the mission of an organisation. Sometimes traits appear because an organisation might be understaffed, or there are too many tasks for each staff member to accomplish in any given day. Therefore a trait may appear that staff begin to employ to save time. Staff may begin to ignore critical protocols to get through their day more quickly. They may then teach these time saving tricks to new staff members. New staff members may begin to abandon other protocols to save more time. Eventually, this negative cultural trait becomes an operational reality of an organisation – and a protocol that, at one point, was developed because it was considered critical, is now ignored. The abandonment of this trait may impact animal welfare, or staff safety.

An organisation, however, has tools at their disposal that can help direct culture in a positive way that enables its mission – while at the same time ensuring that maladaptive traits do not become fixed within the culture. These tools include clearly defined protocols, effective staff training, and ample oversight and accountability.

10.2   A Culture of Safety

Risks and Dangers

Animal care, in general, is a dangerous task. This must be explicitly acknowledged by everyone involved. Whether one is caring for small docile animals, or large aggressive animals, there is always risk involved when caring for another species.[1]

Interactions with other species can always be unpredictable. When dealing with the unpredictability of a species with very strong physical capabilities, this unpredictability can be deadly. Animals, even when habituated to human presence, can exhibit sudden aggressive behaviours – some of which can lead to serious injuries, or even fatalities to care technicians. This risk becomes exponential when dealing with predators such as big cats, powerful primates such as chimpanzees, or extremely large animals such as elephants or whales.

However, it is not just physical attacks that create a risk to animal care technicians. Zoonotic diseases can transfer between animals and humans, posing significant health risks. Additionally, continuous close interaction with animals can cause psychological stress and anxiety for both the animals and the care technicians, potentially leading to safety incidents. Most insidious, however, are the safety risks that have not been predicted.

A **culture of safety** at an animal care facility refers to the collective values, attitudes, and practices that prioritise safety as a fundamental and continuous concern in every aspect of facility operations. Taking the approach of instilling a culture of safety is the best way to prevent even the most unpredictable safety concerns, while mitigating safety incidents that may occur. In essence, a culture of safety integrates proactive safety measures, training, open communication, and responsiveness to create a secure environment for both the animals and the people who care for them, reflecting a commitment to welfare and ethical standards.

All facilities are unique in the risks and hazards that confront staff and animals. Different species, different housing methods, different operating procedures can all contribute to extremely specific safety issues that are unique to the facility. However, many safety issues involve the following categories: containment, the proximity of the managed animals to humans, and basic workplace hazards.

---

[1] See Murphy, 2011.

- *Containment*

  Containment is, obviously, a foundation of animal management. Proper containment is, perhaps, the most critical aspect of safety management in animal care facilities (especially with regard to aggressive, venomous, or very large animals). It ensures the well-being of both the animals and the staff, while also maintaining public safety and regulatory compliance. Proper containment systems prevent escapes, which can be dangerous for both the escaped animals and the public. Enclosures must be designed to withstand the strength and intelligence of the animals they house.

  Containment plays a vital role in disease control within facilities. Proper containment ensures a safe distance between care technicians or visitors, and the animals in their care. Isolation areas are essential for new arrivals to prevent the spread of disease to the resident population. Similarly, quarantine zones are crucial for containing outbreaks of contagious diseases.

  Effective containment is not merely about confinement but involves thoughtful design and management to ensure the safety, health, and well-being of both staff and the animals in their care. Facilities must continuously evaluate and update their containment strategies to adapt to new challenges and improvements in the science of animal welfare. This involves frequent checks on the integrity of all animal containment structures.

  Safety checks on animal enclosures involve a systematic review of structural integrity, containment features, and the suitability of the habitat for the specific needs of the animals housed within. The process includes two key components:

  o *Structural integrity*

    Regular inspections are conducted to assess the condition of fences, walls, gates, and other barriers. This includes checking for wear and tear, potential weak points, and the materials' resistance to the specific behaviours and strengths of different animal species.

  o *Containment features*

    Enclosures must be designed not only to keep animals in but also to keep potential threats out. Safety checks ensure that locking mechanisms are functioning properly, and that there are no gaps or breaches that could lead to escape or entry of unauthorised individuals or other animals.

  o *Double door containment systems*

    Some animals, particularly animals that are prone to quick escapes, or animals that present significant risks to humans, might require a double door containment system. Often referred to as an airlock or sally port, this system consists of two consecutive doors forming a small enclosed space. The primary function of this setup is to create a secure buffer zone that reduces the risk of animals escaping when staff enter or exit animal holding areas.

    The design of double door systems ensures that only one door can be opened at a time. This is typically managed through interlocking mechanisms that prevent the second door from being unlocked until the first door has closed and locked. This setup is crucial in maintaining containment even if an animal manages to push past the first door.

Enhanced safety: These systems are invaluable in enhancing the safety of both the animals and the staff. By preventing direct access to the outside environment, they reduce the stress on animals caused by sudden exposure to unfamiliar spaces and help manage potentially aggressive behaviours triggered by escape attempts.

o *Door and lock checking protocols*

Some facilities employ door and lock checking protocols, whereby all doors and locks are checked by multiple staff members before animals are shifted into an empty area. For example, a facility that houses lions may have three staff members all check the doors and locks before shifting lions into an empty yard. Animals may not be shifted into this area before all three staff members vocalise that they have checked that all doors are closed and all locks are locked.

o *Animal count protocols*

Some facilities may also have protocols for when staff enter an animal area, after the animals have been shifted out of this area. These protocols involve getting an accurate count of all animals within the population to ensure that all animals are out of the area where the care technician is entering. For example, staff may shift lions into an outdoor enclosure. Before they enter the night houses to clean, two other staff members must count all the lions in the outdoor enclosure to ensure that none remain inside.

o *Door and lock maintenance*

Scheduled daily inspections of all doors and locks are critical to containment. These inspections should be carried out by trained staff members who can recognise signs of wear, damage, or tampering. Any issues identified during inspections should be addressed immediately. A system for reporting and prioritising repairs ensures that no faulty door or lock compromises the facility's security.

• *Proximity to humans*

Directly related to containment, but important enough to deserve its own bullet point, is the importance of ensuring a safe proximate distance between the animals in managed care and the humans that care for them or visit them. Each species will have different requirements for such proximity. Often, this proximity goes beyond just the walls of the enclosure. For example, a staff member who gets too close to a mesh wall can still be in potentially grave danger from an animal with extremities that can fit through the mesh.

o *Use of safety lines*

Safety lines are crucial in animal care facilities to ensure a secure distance is maintained between staff and enclosure walls. This practice is vital for preventing accidents and enhancing the safety of both the animals and the caregivers. They are typically marked on the floor or installed as barriers, serve as a physical reminder and a mandatory boundary that staff must not cross during routine or special tasks. These lines are strategically placed based on the specific needs of the enclosure and the species housed within.

Safety lines help in minimising direct contact with animals, which is essential for reducing the risk of provoked attacks or stress-induced reactions from the animals. This is particularly important in enclosures housing large or potentially dangerous species. Additionally, by maintaining a set distance, staff can have a better over-all view of the enclosure, aiding in effective monitoring of animal behaviour and health. This distance also allows for quicker reaction times in emergency situations. The use of safety lines ensures that all interactions with animals are conducted within the safety protocols established by the facility. This compliance is crucial not only for safety but also for meeting ethical standards in animal care.

o *Prevention of zoonosis*

Zoonosis is the spread of disease between humans and non-human animals. It is a critical concern in animal care facilities, where close contact between animals and humans can facilitate the transmission of diseases. Maintaining proper distance is a fundamental strategy to mitigate this risk.

Zoonotic diseases can spread through direct contact, aerosol transmission, or via fomites. In animal care settings, maintaining a safe distance between animals and human caregivers is essential to minimise these transmission pathways.

Implementing physical barriers such as glass panels or fences can help maintain a safe distance without disrupting the care and monitoring of the animals. Clearly marked pathways for staff, separate from animal enclosures, also reduce the risk of accidental close contact.

When close interaction is necessary, personal protective equipment (PPE) such as gloves, masks, and gowns should be used to provide a barrier against pathogens. Different species will require different levels of PPE barriers. These protocols should always come with a recommendation from a veterinary professional.

For animals that are physically handled, staff should be trained in proper handling techniques that emphasise *minimal* contact, and protocols should be in place for dealing with sick animals to prevent the spread of infection to humans and other animals.

o *Bites, grabs, pecks, and so on*

Working with animals presents the obvious risks of being bitten, grabbed, pecked, and so on. These incidents can lead to severe physical injuries, psychological trauma, and in extreme cases, fatal outcomes. These dangers are particularly high in facilities housing large or naturally aggressive species such as big cats, primates, and reptiles. However, even smaller or typically docile animals can pose risks under certain conditions.

The danger of being bitten, pecked, or grabbed in an animal care facility underscores the need for rigorous safety measures and continuous staff training. By understanding the natural behaviours of the animals in one's care, and by preparing for potential dangers, facilities can create a safer environment for both staff and animals.

Bites, grabs, and pecks can result in puncture wounds, lacerations, crushed bones, and in severe cases, amputations. The physical damage depends on the animal's strength, the size of its teeth, beak, or claws, and the duration of the attack.

Bites can also be a vector for zoonoses. Pathogens such as bacteria, viruses, or parasites can be transmitted through saliva, exacerbating the injury with infections that can be challenging to treat.

Encounters that result in being bitten or grabbed can leave lasting psychological scars. Care technicians may develop a fear of the animals they once cared for, leading to anxiety, stress, and even post-traumatic stress disorder (PTSD).

The mitigation of such risks begins with understanding animal behaviour and the specific triggers that might lead to aggressive actions. Facilities must implement comprehensive training programs that educate staff on the signs of distress or aggression in animals and the safest methods for handling them. Staff should receive regular training on the behaviour of the animals they are working with, including how to approach them safely and how to carry out procedures without causing stress or aggression.

Depending on the species, different types of protective gear can be used, such as gloves, face shields, and body armour, to protect against bites and scratches.

Once again, enrichment can be a mitigation tool. Reducing stress in animals through environmental enrichment can significantly decrease the likelihood of aggressive behaviours. This includes providing appropriate enrichment devices, foraging opportunities, and other activities that engage their natural behaviours.

Finally, clear, accessible, and *written* emergency procedures should be in place to handle incidents of bites or other aggressive behaviours. This includes first aid measures and protocols for reporting and responding to such incidents.

- *Basic workplace hazards*

All workplaces contain hazards. However, animal care facilities present some unique hazards that can be extremely significant. However, prevention of these hazards can be achieved through proper preparation and planning.

  o *Trip hazards*

  Trip hazards can vary widely but typically include obstacles that are commonly found in the pathways and work areas of animal care facilities. Here are some specific examples:

  - Substrate: Substrates such as bedding, straw, or wood shavings are essential for animal comfort but can easily be scattered outside of enclosures. This creates uneven surfaces or slippery areas, especially when mixed with water, increasing the risk of slips and falls.

  - Hoses: Water hoses used for cleaning enclosures or providing water can become significant trip hazards if left unattended across walkways. Their flexibility and often indistinct colour make them hard to notice until stepped on, which can cause sudden falls.

  - Tools: Tools such as rakes, shovels, and brooms are indispensable for maintenance tasks but can become hazards if left improperly stored. Placed across walkways or near doorways, they pose a risk to staff moving quickly through the facility.

  To prevent accidents, facilities should implement strict housekeeping rules, such as proper storage of tools and regular cleaning of substrates from walkways. Additionally, installing adequate signage and ensuring that hoses are either brightly coloured or properly stored when not in use can further reduce the risk of tripping.

o *Slip hazards*

Slip hazards are a critical safety concern in animal care facilities, where the combination of animal activity, cleaning processes, and various types of flooring can significantly increase the risk of accidents. They can arise from several sources:

- Wet floors: Water from cleaning, spills from animal drinking areas, or urine can make floors slippery. Facilities often have surfaces that do not adequately resist water, increasing slip risks.
- Uneven surfaces: Wear and tear or poorly maintained floors can lead to uneven surfaces. Transition areas between different types of flooring can also present slip hazards if not properly managed.
- Contaminants: Substrates used in animal enclosures, such as hay, sawdust, or sand, can be tracked into walkways, reducing traction and increasing the likelihood of slips.
- Cleaning products: The use of detergents and disinfectants can leave residues that make floors slippery, especially if not properly rinsed or if used in excess.

Regular checks and maintenance of flooring to ensure there are no cracks, uneven areas, or worn-out surfaces that could contribute to slips. Installation of non-slip flooring materials in high-risk areas, such as wash zones and near water sources, can help reduce these risks. Also ensuring that all staff are trained on correct cleaning techniques, including the proper dilution and use of cleaning agents and thorough drying of floors after washing can be critical.

Emergency Responses

All facilities must have written responses to potential emergencies that might arise. This is the facility's emergency response manual (or ERM). In some cases, oversight bodies may require an ERM, and may even inspect a facility's ERM. Some of the basic sections of an ERM may include:

- Emergency Codes and Responses (for example, standard radio calls that indicate an emergency).
- Responses to animal escapes.
- Responses to fires.
- Responses to dangerous weather.
- Responses to staff/visitor accidents or medical emergencies.
- Responses to animal veterinary emergencies.
- Responses to intruders.

Again, all facilities have specific and unique safety risks, and emergency responses should be tailored to these uniquities.

After-Action Reviews

After-action reviews (or AARs) are structured review processes designed to improve performance by reflecting on the actions taken during a particular event or operation.

For example, if there was an escaped animal and an emergency response protocol were enacted, an AAR would be valuable to perform after the emergency response to determine what went right, what went wrong, what should be done the same, and what should be done differently next time. These reviews are crucial in environments, such as animal care facilities, where learning from experience directly contributes to an organisation's success and safety.

After-action reviews aim to foster an environment of continuous learning and development. By dissecting what happened, why it happened, and how it can be done better, organisations can evolve and enhance their operational effectiveness. Regularly conducting AARs can strengthen team cohesion and communication. They encourage open dialogue and a shared understanding of objectives and challenges. Identifying failures and successes in AARs helps in pinpointing risk factors and developing strategies to mitigate them in future operations.

A truly effective AAR has six main components: preparation, inclusivity, facilitation, analysis, documentation, and the implementation of the lessons learnt.

- Preparation: This involves the collection of all relevant data, including objective metrics and subjective experiences from participants. Preparation might also include setting the agenda and defining the scope of the review.
- Inclusivity: All team members, regardless of rank or role, should be encouraged to share their insights and perspectives. This inclusivity helps in capturing a comprehensive view of the event.
- Facilitation: An effective facilitator guides the discussion neutrally, ensuring that the session remains focused and productive. The facilitator should encourage honesty and prevent the assignment of blame.
- Analysis: The core of the AAR is where the team critically assesses the discrepancies between planned outcomes and actual outcomes, exploring both successes and failures.
- Documentation: Key insights and lessons learnt are documented systematically. This documentation serves as a reference to guide future actions and decisions.
- Implementation of lessons learnt: The effectiveness of an AAR is measured by how well the insights gained are integrated into organisational practices. This might involve updating training programs, revising operational protocols, or enhancing communication channels.

Creating a Culture of Safety

A culture of safety is one in which safety protocols, general safety awareness, a commitment to safe practices, and a common respect for the inherent risks in caring for animals in artificial environments, is a fixed trait within the organisation. Performing tasks in a safe and thoughtful manner is the *only* way of performing tasks. As with all cultural traits, this is passed down to successive generations of care technicians, volunteers, visitors, and anyone else involved with the organisation. This culture is built on a foundation of training, communication, and a shared commitment to safety practices.

The cornerstone of a culture of safety is thorough training. New employees should undergo detailed orientation sessions that cover all safety protocols, emergency procedures, and the correct use of personal protective equipment. Regular refresher courses ensure that safety remains at the forefront of staff activities and helps integrate safety into daily routines. Effective communication channels are vital. This includes regular safety meetings, accessible reporting systems for hazards or incidents, and clear signage throughout the facility. Open communication encourages a proactive approach to safety and ensures that all staff members are aware of ongoing and new safety issues.

Leaders in the facility must exemplify this culture of safety. Leadership stands out as a beacon of a way of doing things that others follow. As such, they should always imbue the culture of safety. This involves not only enforcing safety protocols but also demonstrating a commitment to safety in every action. Accountability measures should be in place to ensure that safety practices are followed, with consequences for non-compliance and recognition for exemplary safety behaviour.

A culture of safety is dynamic, fluid, and evolves with new challenges and information. Regular audits and reviews of safety incidents provide data that can be used to improve safety protocols. Staff should be encouraged to contribute ideas for safety enhancements, fostering a sense of ownership and responsibility.

10.3   A Culture of Respect

In addition to safety risks, assuming responsibility for the care and welfare of another living being can be a deeply emotionally taxing endeavour. One can become so immersed in what they feel is the best for the animals in their care, that they become unbending and untrusting to any other way of doing things, and similarly unbending and untrusting towards other staff members. Similarly, when one person does not fulfil their responsibilities, it can exacerbate and already tense situation. The results can be a culture of intimidation, bullying, resentment, and inconsistency of care. Within such cultures there can exist extremely toxic and harmful behaviours – such as gossip, harassment of co-workers, and even mistreatment of the animals. Ultimately, not only are the staff affected, but the general care and welfare of the animals.

As we have discussed quite a bit in these pages, training and written protocols are extremely effective anecdotes to inconsistency. However, when a culture of respect does not exist, training and written protocols become secondary to such toxicity. Therefore, inherent in the culture of an organisation, must be a culture of respect. All organisations should establish a written and binding code of conduct, and make it a practice to assume positive intent. These practices should be adhered to as policies and governed as such.

A Code of Conduct

A code of conduct in an animal care facility is a foundational document that outlines the ethical standards and professional behaviour expected from its staff. It exists as a written document that clearly defines how all staff are expected to behave in and around the facility, not limited to:

- The treatment of the animals.
- The treatment of other staff members.
- Definitions of harassment, bullying, gossip, or any other inappropriate workplace behaviours.
- How positive conduct enables the mission of the organisation.
- Defined consequences for violating this code.

Establishing such a code is crucial for maintaining a high standard of care, ensuring workplace harmony, and fostering a culture of respect and responsibility. By setting clear expectations for behaviour, a code of conduct helps in cultivating a professional environment. It serves as a reference point for staff, helping them to understand their roles and responsibilities clearly, which is essential for effective teamwork and service delivery. Ultimately, how staff treat each other reigns down to how staff provide for the welfare of the animals in their care. It is a critical component to positive animal care and welfare.

A well-defined code of conduct acts as a deterrent to unethical behaviour by outlining the consequences of such actions. It helps in pre-empting potential issues related to animal welfare, staff interactions, and facility operations. The creation of this code is not merely a bureaucratic step but a strategic measure that enhances the quality of care, supports staff in their professional conduct, and safeguards the facility's ethical standing. It is an indispensable tool for any facility committed to excellence in animal care and organisational integrity.

Assuming Positive Intent

The assumption of positive intent involves approaching interactions and decisions with the belief that all parties are acting with the best intentions towards the welfare of the animals. It fosters a supportive and non-judgemental environment. Assuming positive intent in the context of an animal care facility is a philosophical and operational stance that influences the culture, communication, and care practices within the organisation. This approach can significantly enhance the psychological well-being of both the animals and the staff.

Assuming positive intent can lead to more open communication, reduced workplace conflicts, and a more collaborative atmosphere. It encourages staff to share ideas and concerns without fear of blame, which is crucial for continuous improvement and innovation in animal care.

Basically, this approach works in the following manner: New staff are oriented not just on the technical aspects of animal care but also on the facility's ethos of trust and positivity. This helps instil a mindset of looking for constructive outcomes in every situation. This becomes part of the culture of the organisation.

When issues arise, the focus is on understanding perspectives and finding solutions rather than assigning blame. This approach helps maintain morale and focus on the primary goal of animal welfare. Regular feedback mechanisms that are framed positively can reinforce this mindset. For example, instead of pointing out what went wrong, feedback can be structured around how future situations can be improved.

Assuming positive intent is not just an abstract value but a practical approach that enhances operational efficiency and workplace harmony in animal care facilities. By fostering an environment where staff feel valued and trusted, facilities can ensure that their primary focus remains on providing the highest standard of care to the animals they serve. This philosophy supports not only better outcomes for the animals but also a more fulfilling work environment for the care technicians.

10.4   A Culture of Lucidity

Animals in manage care, as an industry, can obviously be an extremely sensitive enterprise. There is a massive array of opinions and philosophies as to how animals should be cared for. There are a lot of opinions on when and if animals should ever be in managed care. There are a lot of opinions on how animals are used by humans. In fact, many of the readers of this book may have wildly different opinions on these matters. As such, animal care organisations may all proceed under different philosophies of care. This can cause doubts as to the level of care the animals receive that can exist within the general public, and even exist within the staff of an organisation – especially if this organisation is perceived as being secretive about its operations. Secrecy in the care and welfare of animals can cause extremely serious, and sometimes existential, challenges to any animal care organisation. However, the best mitigation strategy to deal with public and staff doubts as to the level of care that animals are receiving is to be completely open, honest, and transparent about the ongoings at a facility – the good things and the challenging things. In short, adopting a culture of lucidity is always a good strategy.[2]

A policy of complete transparency would include the following:

- Defining and articulating the mission of the organisation.
- Being transparent about mistakes and explaining how they occurred.
- Being transparent about animal health issues, including end of life decisions.
- Being transparent about general operations and protocols.

A Clearly Defined Mission

Let's imagine that we run a sanctuary for sea turtles. We might think that just saying 'we have a sanctuary for sea turtles' is enough to define what we do and how we do it. It actually doesn't. Rather we need a clearly defined and articulated mission. This allows staff, supporters, or any other stakeholders the ability to understand the 'how' and 'why' of our activities.

---

[2] See Knowling, 2023.

A clearly defined mission in an animal care organisation is not just a statement of intent; it is the foundational blueprint that guides every aspect of the organisation's operations, from daily activities to long-term goals. When the mission is clearly communicated, all staff members have a uniform understanding of the organisation's goals and methods. This clarity helps prevent misunderstandings about the organisation's practices and policies, which can arise from assumptions or incomplete information. Rather, a well-communicated mission helps staff understand the rationale behind leadership decisions, particularly in challenging situations such as resource allocation or euthanasia decisions. When staff members see how these decisions align with the organisation's mission, it can reduce feelings of resentment and promote a sense of fairness and purpose.

In the absence of clear, transparent communication of how decisions contribute to the overall mission of the organisation, staff may speculate about the motives behind certain decisions, potentially leading to 'conspiracy theories'. This speculation can spread through the organisation and have dire consequences to the general operations and, as such, the care and welfare of the animals.

A mission statement might look something like this:

Lighthouse Turtle Sanctuary is committed to providing rehabilitation for injured sea turtles and lifetime care for sea turtles who cannot be returned to the wild in a habitat where they can thrive and live life to the fullest.

Such a statement can answer questions about decisions regarding:

- Veterinary care: At lighthouse sanctuary, we have to make veterinary decisions on whether or not a turtle can be returned to the wild based upon their condition.
- Husbandry: At lighthouse sanctuary, we provide an appropriate habitat where sea turtles can live like sea turtles. This means having large enough tanks for sea turtles to swim and sandy land areas where sea turtles can bask. It also mean having appropriate climate control, dietary measures, and a solace from human harassment.
- Budget: Taking care of sea turtles is expensive and they live a long time. If we are committed to providing lifetime care, we must make budgetary decisions that ensure our financial solvency.

A well-articulated mission statement serves as a compass for the organisation, ensuring that all efforts are aligned with the core objectives. For staff, this clarity is indispensable. It informs their training programs, helping to instil a sense of purpose and direction. When staff are thoroughly trained on the mission, they are more likely to perform their roles with conviction and consistency, which is essential in maintaining high standards of animal care and welfare.

A defined mission enables the organisation to set measurable goals, which are crucial for evaluating impact and efficiency. This accountability is important not only for internal assessments but also for external credibility. Regulatory bodies, potential partners, and supporters often assess an organisation's legitimacy and effectiveness based on its ability to articulate and adhere to its mission.

10.5   Compassion Fatigue

Compassion fatigue is the physical and/or mental exhaustion, coupled with an emotional withdrawal that affects people in caregiving roles. It results from internalising an individual's perceived agony. It affects all manner of roles within caregiving fields. Animal care is one such role. It is critically important to understand what compassion fatigue is and how it can affect the work a care technician does, the operations of an organisation, and even the general welfare of the animals in the care of the organisation.[3]

First of all, the term compassion fatigue is often poorly understood and incorrectly defined. Sometimes people imagine compassion fatigue as someone saying 'I just care so much, I cannot go on!'. This is NOT what compassion fatigue is. Rather it has to do with the anxiety one feels while in a caregiving role based on the immense responsibility one feels in being in charge of critical elements of an individual's life and well-being.

Actually, compassion fatigue is often subconscious. In fact compassion fatigue can affect your emotions to actually make you feel less compassionate! The effects of compassion fatigue can cause someone to withdraw to the point where they no longer feel a connection with much of the world around them – including the animals in their care.

Compassion fatigue is different from burnout. Burnout can be experienced in any occupation (for example, an airline counter worker may get burnt out when dealing with angry customers). However, compassion fatigue refers directly to those in caregiver roles that may internalise perceived suffering (doctors, nurses, first responders, animal care technicians, etc.).

According to the Compassion Fatigue Awareness Project, some signs of compassion fatigue are the following[4]:

- Excessive blaming.
- Bottled up emotions.
- Isolation from others.
- Substance abuse used to mask feelings.
- Compulsive behaviours.
- Poor self-care (i.e., hygiene, appearance).
- Voices excessive complaints about administrative functions.
- Legal problems and indebtedness.
- Nightmares and flashbacks to traumatic event.
- Chronic physical ailments such as gastrointestinal problems and recurrent colds.
- Apathy, sad, no longer finds activities pleasurable.
- Mentally and physically tired.
- Being constantly preoccupied.

Within an organisation, compassion fatigue can lead to the following:

- High absenteeism.
- Constant changes in co-worker relationships.

---

[3] See Bride et al., 2007; Sinclair et al., 2017; and Stoewen, 2020.
[4] See Compassion Fatigue Awareness Project.

- Inability for teams to work well together.
- Desire among staff members to break company rules.
- Outbreaks of aggressive behaviours among staff.
- Inability of staff to complete assignments and tasks.
- Inability of staff to respect and meet deadlines.
- Lack of flexibility among staff members.
- Negativism towards management.
- Strong reluctance towards change.
- Inability of staff to believe improvement is possible.
- Lack of a vision for the future.

What Can One Do to Combat Compassion Fatigue?

On an individual level, the best defence against compassion fatigue is acknowledging that it exists. One can do this by recognising the signs and symptoms and understanding that when it occurs, it is important to communicate how one is feeling to co-workers and supervisors.

Additionally, it is important that an individual can both physically and mentally separate from the organisation where they work. To do this, one must that the facility can exist without them when they are not present. This means 'mental leaving' work at the end of a shift. It also means being able to say 'no' if asked to take on an additional duty or shift if one is feeling stressed. Care technicians should take any paid time off they may have accrued. Similarly, no one should succumb to guilt for taking a day off. Care technicians should engage in hobbies that are not related to the work they do. In fact, it is very healthy to have projects unrelated to animal care completely!

It is important for a care technician to 'know their exits'. This means understanding that one can call for relief during a difficult situation at work. It means having the ability to reach out to supervisors and co-workers for assistance. It also means taking breaks when appropriate.

Finally, a care technician should also trust the animals. This means understanding that the animals have survived without them in the past and will survive without them in the future. This also means understanding that there are many times of the day that the animals exist without direct care – and survive.[5]

How Can an Organisation Adopt a Culture of Compassion Fatigue Awareness?

To combat compassion fatigue, organisations would do well to adopt an 'E.R. Model', in that the organisation functions as a collection of emergency room doctors and nurses. This means encouraging staff to deal with the emergencies when they arrive on property but encouraging them to physically and mentally go home when they are relieved.

[5] See Schwanz & Paiva, 2022.

Organisations should discourage all aspects of a 'blame culture' – where staff always seek to find an individual responsible for whatever mishap might occur. Accountability should always be handled in a professional manner by the appropriate staff members – never through an onslaught of coworker blame. Blaming and second-guessing the actions of co-workers after the fact should be immediately rejected as the inappropriate banter it is.

Organisations must also foster trust within its staff. Staff have to be able to trust the organisation to do its best in following its mission. Staff should never have a reason to assume the worst. This means that organisations must transparently evolve where necessary while retaining its best elements. Inherent in this is accepting accountability on an organisational level, and admitting when mistakes are made.

Organisations should become networks of support. The organisation should ensure that work is a safe environment where staff can communicate their challenges without fear of gossip or innuendo. Organisations should ensure that work is an environment where relief is granted when needed. Organisations should ensure that it is fostering an environment where coaching and support is offered rather than blame.[6]

Things to Remember about Compassion Fatigue

- Staff are not defined by their sacrifices.
- The fact that the staff work in an animal care capacity means that they care very deeply for the animals and, likely, the mission of the organisation.
- Care technicians are entitled and deserve a life outside of the organisation and its residents.

Implications

The culture of an animal care organisation is deeply rooted in a commitment to the welfare and ethical treatment of animals, underpinned by a framework of safety, respect, transparency, and continuous improvement. Safety is paramount, with protocols and training in place to protect both animals and caregivers from the inherent risks associated with animal care. Respect is crucial, not only towards the animals but also among staff members, fostering an environment free from bullying and harassment, and where every individual feels valued. Transparency is actively practised, ensuring that all operations and decisions are open to scrutiny by all stakeholders, which helps in building trust and accountability.

Additionally, the organisation must recognise the emotional toll that intensive caregiving can have on staff, addressing compassion fatigue with supportive measures and resources to ensure mental well-being.

[6] See Paiva & Schwanz, 2022.

Lastly, a culture of continuous learning is encouraged, where staff are motivated to engage in ongoing education and feedback mechanisms, allowing for the evolution of care practices based on the latest knowledge and experiences. This holistic approach ensures that the organisation not only meets the physical needs of the animals but also supports the psychological well-being of both the animals and the staff, creating a harmonious and effective working environment.

References

Bride, B. E., Radey, M. and Figley, C. R., 2007. Measuring compassion fatigue. *Clinical Social Work Journal*, *35*, pp. 155–163.

Compassion Fatigue Awareness Project. Available at: https://compassionfatigue.org/ (Accessed: 26 November 2024).

Knowling, P., 2023. The view from beyond the fence: ageing zoo animals and communicating with the outside world. In *Optimal wellbeing of ageing wild animals in human care* (pp. 253–268). Cham: Springer International Publishing.

Murphy, H. W., 2011. Dangerous animal crisis management. *Fowler's Zoo and Wild Animal Medicine Current Therapy*, *7*, p. 78.

Paiva-Salisbury, M. L. and Schwanz, K. A., 2022. Building compassion fatigue resilience: awareness, prevention, and intervention for pre-professionals and current practitioners. *Journal of Health Service Psychology*, *48*(1), pp. 39–46.

Schwanz, K. A. and Paiva-Salisbury, M., 2022. Before they crash and burn (out): a compassion fatigue resilience model. *Journal of Wellness*, *3*(3), p. 7.

Sinclair, S., Raffin-Bouchal, S., Venturato, L., Mijovic-Kondejewski, J. and Smith-MacDonald, L., 2017. Compassion fatigue: a meta-narrative review of the healthcare literature. *International Journal of Nursing Studies*, *69*, pp. 9–24.

Stoewen, D. L., 2020. Moving from compassion fatigue to compassion resilience Part 4: Signs and consequences of compassion fatigue. *The Canadian Veterinary Journal*, *61*(11), p. 1207.

11　Pitfalls

Figure 11.1 A cougar at the Memphis Zoo. (Photo credit: author).

11.1   The Juniper Tree

A tear falls down The Mariner's face. The saltiness of it keeps it from freezing as it travels down his cheek and into the forest of his snowy beard. He wipes his eyes as he looks at the juniper tree in front of him. He turns around to find his companion. In the haziness of the white frozen landscape, he can only make out the shape of his travelling mate.

'This is it!', The Mariner exclaims.

The Mariner marvels at the green colour amidst the backdrop of white. He touches the branches and smells the scent of evergreen. Deep in the branches, he sees the berries he has been seeking. He picks one and puts it in his mouth. The bitterness makes him wince. Still, he bites down on it. He smiles.

He hands one to his companion. The companion also eats the berry. The two of them look around and survey the forest of these trees. It has been a very long journey to finally get here. Fortunately, they were able to find the island. The myths of dangerous whales guarding the island had been just that – myths.

Picking up one of the burlap sacks that they had dragged across the ice, The Mariner instructs his companion to help him pick enough berries to fill each. The two men begin to pick berries from each tree. When they gather as much as they can carry, they begin to drag the bags back towards their rowboat on the icy shore of the island.

When they reach their rowboat, they look out at their ship anchored a short distance away in the icy sea. They put the bags in the boat and row out towards the ship. As they near it, they hear a very low rumble. The rumble seems to be coming from the depths of the sea. Waves begin to rock their rowboat. A large wave tips them over, submerging them in the freezing cold water. The burlap sacks of berries fall into the sea. The two men begin to try to swim to their ship. Out of the icy white water emerges an enormous black bowhead whale. The whale breaches the water and lands with a splash. The splash creates a wave, which carries the two men away from their ship and back towards the island.

Freezing, the two men crawl on to the shore. They look back at their ship, only to see the giant whale's tail emerge from the water and crash down upon the bow. Splinters of wood fly into the air. The two men gasp as they see their ship fill with water and begin to sink into the sea.

The cold begins to overtake both men. Almost simultaneously, they begin to lose consciousness. They are not aware of the help that has arrived.

A hermit, an 'iceman', who lives on the island has found the two men. He places both on a sled and drags them across the iced earth. Across the windy white plains, The Iceman travels quickly. He hears the low rumble of the whales from the sea. This makes him move even faster. Finally, he arrives at the black rocks. He lifts each man from the sled and pulls them into the crevices of the rocks. Through the crevices, he lowers them into the black cavern where he makes his home.

As he drags them lower into the cavern, the air grows warmer. Finally, they reach a large room with fire pits, furniture, and tapestries on the walls. The warmth of the environment causes the two men to slowly regain consciousness. As they stir, The

Iceman pushes several berries into their mouths. The bitter taste of the berries begins to stir them awake.

The Mariner awakens first. He sees the beautifully adorned room and feels the warmth of the air.

'Have I died?' he asks The Iceman.

'No,' says The Iceman, 'this is my home. And now, as sad as you may find this, it is your home too.'

The Mariner's companion begins to awaken as he chews on the berries in his mouth. He looks around. He doesn't look alarmed. He closes his eyes again.

'The whales keep us here,' says The Iceman. 'We are their pets, if you will. They watch us. They never let us leave. Once you are here, you are here for life.'

'There are several islands just like this one. On rare days, you can see the other islands across the water. Each island is inhabited by stranded mariners. Each mariner arrived here in search of the trees.'

'We can never leave. We can't even really ever travel to the neighbouring islands. A few have done it before, but it's not really worth the risk. If you attempt it, the whales will stop you. Some have constructed rafts. However, when they've left the islands, the whales have surrounded them and forced them back. On a few occasions, the whales have drowned those who have gotten too far away from the islands.'

'I am the only one left on this island. Years ago there were others. One of them was very old. He ended up getting very ill and died. The other attempted to escape. The last I saw him, he faded into the hazy mist as he drifted away on the raft he constructed. As soon as he left, I heard the whales and saw their tails. I imagine he either ended up on another island, or the whales were forced to drown him.'

The Mariner shook his head in disbelief. 'There must be a way off the island! I am on a very important quest.'

The Iceman shook his head. 'I don't think you understand. Whatever 'quest' you were on no longer exists. There is *no* way to leave. Everything your life was before landing here, is no longer. Whatever you owned, whoever you knew, whoever you hated, whoever you loved, and all of your 'quests', no longer matter. The only thing that exists for you is this world – this environment. Your only quest is to find a way to adapt to it.'

'So all of this has been for nought?' asked The Mariner.

'Anything that preceded this moment,' replied The Iceman, 'only matters insofar as the fact that these events brought you here.'

'Why do the whales keep us here?' cried The Mariner.

'As I said, they watch us. They study us. They learn from us. But, honestly, it really isn't that bad. I have been here so long that I do not remember my life before. I'm not even sure where I was born. Maybe I was born here. It really doesn't matter. The only thing I know about the outside world, or even the world on the other islands, comes from those that have been on this island with me before. I live a fine life. I paint my tapestries. I am well fed with the crustaceans under the ice. When I get sick, the berries from the trees make me feel better.'

'You will soon accept things as they are. This world around you has everything you need to survive. In fact, it has even more than that. It has everything you need to live life to its fullest potential.'

The Mariner buries his face in his hands. He begins to weep. 'I have come such a long way. I have gone through so much. I just want to complete this mission and get back to my life.'

The Iceman raises one eyebrow and gives a sympathetic smile. 'Your mission is over. Your old life is over. You have now have a clean slate. This is a new world for you to adapt to. This is your new environment. You'll have a new life and you'll adapt to it. You'll find a way to thrive in this world of ice – a world far removed from where you came and from where you had adapted.'

Later that evening, The Mariner's companion awakens. He bundles up in The Iceman's clothing and goes outside. He exits through the crevice in the rock outcropping. He feels the cold air on his face as he walks across the frozen earth. He continues until he reaches the shore of the island. He notices how clear the night sky is. The stars shine above. The aurora glistens like colourful curtains of light. Everything is quiet but the rumble of whales. Everything is still except for the ripples on the water that the whales create. He sees the spray from their blowholes as they rise to the surface. He wades out into the water. All at once he feels the connection to his old life. His past as a captain, castaway, and mariner's companion, all fade into the distance of his old worlds. The distance feels immense.

The freezing water begins to numb his legs. He turns back towards the shore. He will face this new life. He has companions. He has a warm home. He has food. He has berries to make him better when he is sick. He will survive. He will find a way to live life to its fullest potential. The whales will allow him the freedom to do just that.

He walks back on to the shore. He heads back to the black rock outcropping. He will take comfort in his new night house. He will awaken tomorrow, ready to find a life to live in the whales' menagerie.

Could We Thrive?

What would we need in such a situation? How could we possibly thrive in a world where the climate is so different from what we are acclimated to? How could we manoeuvre about in a world where the terrain is not what our bodies were adapted to? How could we deal with being watched all the time by another species? How could we sustain an existence where we understood that, no matter how good we might have it in our habitat, we can never actually leave it?

One thing we would likely need is for those taking care of us to be aware that, no matter how much they may make our artificial habitat liveable, it is still a *managed* environment. As such, they would need to take every scientific approach possible to giving us the freedom to thrive in such a situation. This means that their job would be to provide for us, rather than treat us like their pets. It means that if they made a mistake with our care, they would be transparent about it and accountable for it so that they could receive assistance and guidance. It means that they would work positively with each other so that they could be provide us with a standard of care. It means that they would find the resources required in order to meet our care needs. It means that their mission, above all, would be our care. In short, it means that they would avoid the standard pitfalls that frequently hamper care technicians.

What Are 'Pitfalls'?

Pitfalls in animal care are the operational challenges that make it more difficult to provide a place where animals can thrive. They frequently occur when an inferior aspect of animal care takes priority over the mission of allowing animals to thrive. This can mean caring for animals in an improper way – such as treating them like pets rather than individuals in need of care and in need of a maintained habitat that allows and promotes thriving. This can mean operating in secrecy due to the fear of public reactions or staff reactions to circumstances that might come up in an animal care facility This can mean entering into a long and protracted battle with regulatory agencies – where fighting compliance issues overtakes the science of providing care. This can mean running into budgetary issues that cause an organisation to cut aspects of care that allow for thriving. This can mean entering into rivalries with other organisations – ensuring that the rivalry takes precedence over allowing and promoting animals to thrive.

All of these pitfalls can be easily avoided. To do so, facilities must be aware of them and recognise them when they arise. They should be seen as anathema to proper care. As we will see, like many other aspects of animal care and welfare, pitfalls can be avoided by using a scientific bedrock, a philosophy of care, which is a commitment to the welfare and thriving of each resident animal.

11.2   Pitfall 1: The Pet Mentality

I once had dealings with an animal sanctuary where the animals were frequently given special 'treats' (candy, cheesecake, breakfast pastries, and several other junk food items that could have severe health consequences for animals living in managed care). When I protested about this, I was told that 'these animals have been through so much in their lives that we like to spoil them here'. What was especially rich was that this sanctuary housed mostly obese animals – some of which even had diabetes. I was puzzled over the fact that 'spoiling' these animals took precedence over providing them the proper care they required to lead a healthy life.

I continued to investigate the situation. What I found was that the staff seemed to be gratified by the reaction they were getting from the animals when they would give them junk food. This reaction was, in fact, so gratifying to the staff (and leadership), that receiving the gratification was more important than the health of these animals. As a result, the animals were living in a state of poor health – a state entirely created by this practice.

Sometimes it is difficult to treat the animals in our charge as the patients of care that they are, rather than our pets (or, in some cases, like children). As we keep finding, it is always important to realise that the animals in our care are individuals with individual needs, challenges, preferences, health statuses, anxieties, and so on, and it is our role to scientifically meet these needs. There are often times that meeting these needs comes with little to no gratification. There are also times that meeting these needs results in losing the trust of an animal or causing unavoidable discomfort or even pain. However, our role as care technicians it to meet these needs for the welfare of that individual.

This is actually a common pitfall that can plague even the most seasoned care technician. It is what I term the 'pet mentality'. This mentality can be present when a care technician seeks gratification from the animal, or animals, in their care, and this gratification takes precedence over the animal's welfare. For example, if one brings a slice of cheesecake to a chimpanzee, the chimpanzee will likely vocalise happy food grunts. The care technician may feel gratified that they have made the chimpanzee happy. However, the slice of cheesecake contains all sorts of ingredients that a chimpanzee should not be consuming.

The pet mentality can also rear its ugly head when a care technician wants to spare one of the animals from the discomfort of an essential medical procedure. For example, an animal may require restraints so as to not pick at a surgical suture sight. A care technician may be very troubled seeing the distress that the restraints cause. Instead of finding a way to minimise this discomfort, while still restraining the animal from picking at the suture site, a care technician may remove the restraints altogether. This could have catastrophic results.

The pet mentality takes on even greater significance if it plagues an entire facility or organisation. This can happen when leadership is not well educated in standards of care. Though well meaning, the organisation can become a place of deep animal welfare challenges.

This is why it is critical to identify the pet mentality and address it immediately. Leadership of facilities should consult with experts in the field. Boards of Directors should also meet regularly with experts to better govern a facility.

11.3   Pitfall 2: A Lack of Transparency

It is very well documented that transparency is an essential aspect of an overall mission of any organisation that engages in animal care (see Chapter 4.1). In fact, it has been shown that transparent organisations have a much greater buy in with staff and stakeholders regarding their overall mission and objectives.[1] When a lack of transparency is detected, the overall mission suffers. When this mission is to find scientific methods for animals to thrive, then a lack of transparency can directly create animal welfare issues.

When transparency is lacking, there can be a disconnect between what the organisation claims to do and what actually happens on the ground. This misalignment can lead to practices that may not prioritise the best interests of the animals, thereby preventing them from thriving.

Transparency is of utmost importance in monitoring and maintaining high standards of animal care. Without transparency, issues such as neglect, inadequate care, or inappropriate treatment may not come to light until they have caused significant harm. This lack of visibility can prevent timely interventions and actually allow poor practices to become entrenched within the culture of the organisation.

First off, when a lack of transparency gets discovered, it destroys all trust between an organisation, its staff, and the public. This trust is vital for the support and sustainability of any animal care facility. Transparency fosters trust, while its absence can lead to suspicion and scepticism among the public, staff, volunteers, and donors. If

[1] See Mills et al., 2018.

stakeholders suspect that the facility is hiding information about animal treatment, financial management, or operational decisions, it can lead to a withdrawal of support, which is crucial for the facility's functioning and expansion.

Transparency impacts internal operations. A transparent management fosters a culture of accountability and ethical behaviour. Conversely, opacity can lead to mistrust and low morale among staff, undermining teamwork and efficiency. It can also stifle innovation and improvement, as staff may be less likely to speak up about concerns or suggest improvements if they feel information is withheld or manipulated.

Financial donors and operational partners prefer to associate with organisations that demonstrate clear and honest communication. A lack of transparency can make it difficult to secure funding or partnerships, limiting the facility's ability to maintain or expand its services. Additionally, undisclosed financial struggles or mismanagement can lead to sudden financial crises, leaving staff and animals in precarious situations.

The long-term success of an animal care facility hinges on its reputation, which is significantly influenced by its level of transparency. Persistent issues with transparency can lead to a decline in visitor numbers, a decrease in donations, and difficulties in attracting and retaining qualified staff. This can create a cycle of negativity and decline, challenging the facility's ability to fulfil its mission.

11.4   Pitfall 3: Battles with Regulatory Bodies

The role of accrediting bodies and regulatory agencies is pivotal in ensuring that organisations adhere to the highest standards of animal welfare and ethical practices. These entities provide essential oversight through a structured framework that includes setting standards, conducting evaluations, and enforcing compliance. However, while they provide guidance, oversight, and establish measurable standards, they can also be a major challenge for animal care organisations. Sometimes this can lead to long and protracted battles with oversight bodies. These battles can completely destroy an organisation.

The Role of Accrediting and Regulatory Bodies

Accrediting bodies and regulatory agencies develop comprehensive guidelines that define the minimum requirements for animal care, facility management, and staff training. These standards are often based on scientific research, best practices in the industry, and ethical considerations. For example, the European Association of Zoos and Aquariums (EAZA) provides detailed accreditation standards that cover aspects ranging from animal health and diet to enclosure design and emergency protocols.

Oversight is operationalised through regular inspections and assessments conducted by the accrediting bodies. These evaluations are designed to ensure that animal care organisations not only meet the established standards at the time of accreditation but continue to maintain these standards over time. Inspectors assess various facets of the organisation, including animal health, living conditions, and the effectiveness of the existing management systems. Feedback from these evaluations helps organisations identify areas for improvement and implement corrective actions.

When discrepancies or violations are identified, regulatory agencies have the authority to enforce compliance through various means. This might include issuing fines, mandating specific changes to operations, or, in extreme cases, revoking accreditation. Enforcement ensures that organisations take the necessary steps to align with best practices and legal requirements, thereby safeguarding animal welfare.

Through these mechanisms, accrediting bodies and regulatory agencies play a crucial role in promoting high standards of care and ethical treatment in animal care organisations. Their oversight ensures that the welfare of the animals is prioritised and that the organisations operate transparently and responsibly.

Long and Protracted Battles

When an animal care organisation becomes embroiled in a protracted battle with an oversight or regulatory agency, the conflict typically stems from several core issues: compliance failures, communication breakdowns, and divergent interpretations of regulatory standards.

* *Compliance failures*

    The most direct path to sustained conflict is through non-compliance with legal and ethical standards set forth by oversight bodies. For example, an organisation might fail to meet the housing standards for animals, such as providing adequate space for elephants as mandated by AZA. If these standards are not met, and the organisation does not rectify these conditions promptly after they are pointed out by inspectors, it can lead to fines, ongoing legal disputes, or more severe penalties.

    Effective communication is essential in maintaining a cooperative relationship with regulatory agencies. A lack of clarity or transparency in the organisation's reporting can lead to misunderstandings. For instance, if an animal care organisation does not fully disclose an incident where an animal was harmed due to staff negligence, and this comes to light through other means, it can damage trust and lead to increased scrutiny and prolonged disputes.

* *Divergent interpretation of standards*

    Conflicts often arise from differing interpretations of what compliance means. An organisation might believe its practices are compliant and in the best interest of animal welfare, while a regulatory body may view these practices as inadequate. For example, an organisation might use tethering as a control method in elephant management, which it considers necessary for safety and control, while inspectors might see this as a violation of the animals' natural behaviours and welfare, leading to extended audits and reviews.

    Regulatory agencies and animal care facilities can have all sorts of wildly divergent interpretations of standards due to several factors, including differing priorities, varying levels of expertise, and the ambiguity of regulatory language.

Regulatory agencies are primarily concerned with ensuring compliance with laws and regulations designed to protect animal welfare and public safety. Their approach is often more conservative, prioritising strict adherence to rules and guidelines. In contrast, animal care facilities may prioritise practical aspects of animal management, focusing on day-to-day operations and the immediate well-being of the animals. This difference in focus can lead to conflicting interpretations of what constitutes adequate care and compliance.

Animal care facilities often employ staff with hands-on experience and practical knowledge of animal behaviour and care. These professionals may have a deep understanding of the specific needs and behaviours of the animals they manage, which might lead them to interpret standards in a way that aligns with their practical experience. Regulatory agencies, on the other hand, might rely more heavily on scientific studies and formal guidelines that may not always capture the nuances of individual animal needs or the latest in situational ethics.

Standards and regulations are sometimes written in language that is open to interpretation. What one party sees as clear and straightforward, another might view as vague or open-ended. For example, a regulation might require that animals be given 'adequate space' without specifying exact dimensions, leading to different interpretations of what 'adequate' means based on the species, the physical condition of the animals, or the design of the facility.

These factors contribute to the potential for significant discrepancies in how standards are understood and applied. To bridge these gaps, ongoing dialogue, joint training sessions, and collaborative reviews of standards can help align the interpretations and ensure that the primary goal of animal welfare is met effectively and efficiently.

Mitigating Differences with Regulatory Bodies

To mitigate these issues, organisations should prioritise transparency, maintain open lines of communication with regulatory agencies, and ensure that they fully understand and comply with all legal requirements. Regular training on compliance, proactive engagement with regulators, and the utilisation of mediation or legal counsel when disputes arise are also critical in managing and resolving conflicts effectively. These strategies help align the organisation's practices with regulatory expectations and foster a cooperative rather than adversarial relationship with oversight agencies.

11.5   Pitfall 4: Austerity Measures That Compromise Welfare

Imagine an animal sanctuary that prides itself in providing a place where animals can live out their lives with exemplary care. The sanctuary relies on donations to exist. However, for whatever reason, donations are lower than expected. The sanctuary goes through their budget to determine what to cut out. The leadership of the sanctuary sees 'animal substrate' occupying a large area of budgeted funds. They investigate its use. In their investigation, they find that staff dislike the substrate because it is heavy and difficult to clean. They find that, in addition to purchasing the substrate, they have to pay for its waste removal. They find that their maintenance staff complain that substrate clogs up the drains of the animal housing facilities. At this, they make what they believe is an easy decision. They cut animal substrate from the budget. Unfortunately, animal wastes do not break down in the pens. Animals are covered in their own waste. As a result, the rate of diseases increases. Additionally, staff now have to actually clean more areas because there is no substrate to absorb the animal wastes. About is increased, costs are increased, and animal welfare has been compromised.

Austerity measures in animal care facilities, often implemented as cost-cutting strategies, can significantly compromise the ability of animals to thrive. These measures can lead to reductions in the quality and quantity of food, veterinary care, and enrichment activities, directly impacting the physical and psychological well-being of the animals.

- *Nutritional deficiencies*

  Cost reductions often result in cheaper, lower-quality food or less frequent feeding schedules. Proper nutrition is fundamental to an animal's health, growth, and vitality. Nutritional deficiencies can obviously lead ton array of health problems.

  Budget cuts often reduce the variety of food offered. This not only affects the physical health of the animals due to a lack of balanced nutrients but can also impact their mental health. Animals in captivity benefit from dietary variety, which stimulates their cognitive functions and mimics their natural feeding behaviours.

  Animals with specific dietary needs due to health issues, age, or breeding requirements may suffer disproportionately when quality and variety of food are compromised. This can lead to exacerbated health problems.

  Cheaper alternatives to high-quality feeds might not meet the specific nutritional requirements of different animal species. For instance, replacing specialised food with lower-cost generic food can result in diets that lack essential nutrients, vitamins, or minerals necessary for the health and well-being of the animals.

  Cost constraints might also lead to irregular supply of food items, disrupting the regular feeding schedules that are crucial for maintaining the health and metabolic stability of animals. This inconsistency can stress animals and lead to health issues related to poor nutrition.

- *Inadequate veterinary care*

  Austerity may lead to less frequent veterinary visits and preventive care, including vaccinations and routine check-ups. This can result in the late diagnosis of diseases, which in advanced stages, might be more difficult, costly, and less likely to be successfully treated. Chronic conditions might go unmanaged, leading to unnecessary suffering.

  Budget cuts often result in fewer staff members, which can lead to increased workloads for existing staff. This can result in rushed consultations, less time for each patient, and potentially overlooked symptoms or conditions.

Essential maintenance of medical equipment and facility upgrades may be postponed due to budget cuts. This can affect the functionality of the veterinary practice and potentially compromise the safety and comfort of both animals and staff.

Overall, budget cuts can severely impact the basic effectiveness of all veterinary services. This can actually lead to higher long-term costs based on untreated conditions (or poorly managed conditions).

- *Reduced enrichment*

  Enrichment is unfortunately sometimes viewed as discretionary and oftentimes suffers from cuts during austerity measures. This, obviously, carries with it some severe consequences in terms of animal welfare. Animals are then not stimulated mentally, physically, or even engaging much with their environment.

- *Staffing shortages*

  Reduced budgets may lead to layoffs or hiring freezes, increasing the workload on remaining staff. This can decrease the quality of care due to less individual attention per animal and increased stress on staff, potentially leading to burnout and high turnover rates.

  Additionally, 'cheaper' staff with less experience or education may be hired. This lack of experience can have profound consequences in carrying out the mission of the organisation.

So How Then Are Animal Care Facilities to Deal with Times of Austerity?

It is critical that austerity measures be researched. This should be done by consulting other facilities who may have made similar cuts, consulting with animal care experts, and with discerning the reason why this was seen as a critical budgetary cost in the first place. A basic rule of thumb is that if something is included in a budget, at one point it was viewed as a critical need. If so, has something changed that makes it less critical now?

Using Dynamic Modelling

Oftentimes the source of such poorly thought out austerity measures lies in a budget that doesn't include the unpredictable nature of animal care. While fluctuations in revenue may be predictable (e.g., fluctuations in attendance at a zoo), fluctuations in expenses may be seen as fixed. However, this is obviously not the case. Animals may develop expensive physical challenges. Facilities may be unexpectedly damaged. In the care of animals, unexpected expenses is usually the rule rather than the exception.

Dynamic modelling is a method of budgeting that takes into account how systems such as sources of revenue and financial needs change over time. Using it can significantly aid an animal care facility in managing its budget by providing a comprehensive tool for forecasting, planning, and decision-making. This approach allows care technicians to simulate various scenarios and their financial outcomes, enabling more informed budgeting decisions.[2]

---

[2] See Chalupova et al., 2014.

Basically, it works like this: an animal care facility looks over time on what it has spent its money on – how much on food, how much on veterinary care, how much on staff wages, and so on. At the same time, it looks at variables that can affect these expenditures. For example, a facility may look at things like animal aging and predict a greater need for veterinary care. It may look at staff performance issues and determine that better training might lessen the need for more staff. It might look at inflation in food costs. Then it should look at how much revenue was generated and what the predictable changes to this revenue might be. For example, maybe travel to an area with a zoo might be forecast to increase, or decrease. Perhaps a major granting source is making cuts. Then a scenario is run, taking all these factors in – at the same time making sure that all of the critical components of thriving are maintained in each scenario.

Dynamic modelling helps predict future financial scenarios based on current data and trends. For instance, it can simulate the impact of changes in visitor numbers, financial donations, in kind donations, variations in funding sources, or fluctuations in operational costs. This predictive capability allows animal care facilities to plan their budgets more effectively, ensuring that they allocate resources where they are most needed to support animal welfare and facility maintenance.

By using dynamic models, animal care facilities can optimise the allocation of their limited resources. The model can highlight areas where spending is critical for maintaining animal health and where there might be opportunities to reduce costs without compromising care standards. This helps in prioritising expenditures that directly contribute to the animal care mission of animal welfare.

Dynamic modelling enables animal care facilities to create and analyse 'what-if' scenarios. For example, they can assess the financial impact of a new exhibit or a conservation program before committing funds. This analysis helps in understanding potential returns on investment and avoiding financially unviable projects.

Dynamic modelling can integrate various performance indicators from different departments within an animal care facility, providing a holistic view of the organisation's financial health. This integration helps identify inefficiencies and areas for improvement, leading to better financial management and operational efficiency.

11.6   Pitfall 5: Rivalries

Competition can be healthy. Competition between animal care facilities can actually lead to better care. It can lead to innovations. It can lead to more efficient methods. It can lead to competing for the most talented staff members. Competition can even lead to better information sharing, if an organisation is doing something really well and wants to use public knowledge of what it's doing as a way to outdo its competitors. However, competition can get out of control. It can lead to rivalries between organisations that end up adversely affecting animal welfare.

The shape and scope of rivalries oftentimes are specific to the types of animal care organisations. For example, non-profit sanctuaries may compete for donors.

Zoological parks may compete for guests. Laboratories may maintain scientific rivalries based on publications. All animal care organisations may compete for grants.

While competition can drive improvements and innovation, it is crucial that animal care facilities prioritise the welfare of the animals above rivalry gains. Ensuring that competition does not come at the expense of animal well-being requires vigilant regulation, a strong ethical framework, and a commitment to the primary mission of animal conservation and care.

Below are some general examples on how rivalries can impact animal welfare.

* *Resource diversion*

  In an effort to outdo competitors, facilities might allocate substantial resources towards marketing and aesthetic improvements that generate publicity, or attract visitors, but do not necessarily enhance animal welfare. This diversion of funds can result in a lack of investment in essential areas such as habitat enrichment, veterinary care, and staff training, directly affecting the quality of life for the animals.
* *Overemphasis on profitability*

  Competitive pressures can lead facilities to prioritise profitability over the welfare of the animals. This might manifest in overcrowding to increase ticket sales, under-staffing to reduce operational costs, or even neglecting the needs of less 'popular' animals that do not draw significant public interest or revenue.
* *Stressful environments*

  The desire to create more engaging or interactive experiences for visitors can lead to the development of environments that are stressful or unsuitable for animals. For example, increased noise levels, frequent night-time events, or excessive handling by visitors can significantly disturb the natural behaviours and stress levels of animals.
* *Ethical compromises*

  In the race to be seen as the most innovative or unique, facilities might make ethical compromises that can harm animals. This could include the acquisition of exotic or high-maintenance species without adequate preparation for their care, or engaging in breeding practices that do not consider the long-term welfare of the animals.
* *Lack of collaboration*

  Rivalries can lead to a lack of willingness to share research, best practices, or resources between facilities, which is detrimental to the overall advancement of animal welfare standards. Collaboration often leads to better outcomes for animal care, and without it, facilities may miss opportunities to improve their care standards or innovate in ways that benefit the animals.

11.7   The Solution: A Philosophy of Care

A **philosophy of care** is a constitution of sorts. Basically, an organisation's written philosophy of care governs all present and future care protocols, thus ensuring the mission and vision of the facility with regards to animal welfare. A philosophy of care is a foundational framework that outlines the core principles, values, and approaches an animal care facility adopts to ensure the well-being and ethical treatment of the animals under its care. This philosophy serves as a guiding document that influences all aspects of animal management, from daily care routines to long-term welfare strategies.

For example, a chimpanzee sanctuary may have a philosophy that states 'we value freedom of choice, self-determination, and a rich social environment for every chimpanzee; therefore all care protocols should prioritise giving chimpanzees as much space, enrichment, and companionship as possible with respect the limits of an individual's social and spatial threshold'. Once this is written in a philosophy of care, all present and future protocols would have to conform to this. As such, a protocol that states that all chimpanzees would be separated into small cages to avoid aggression would not be permissible by the philosophy of care.

A philosophy of care ensures that all staff members are aligned with a unified approach to animal welfare. This consistency is crucial for maintaining high standards of care across all levels of the organisation. By adhering to a well-defined philosophy, facilities can systematically address the physical, psychological, and social needs of the animals. This holistic approach helps in promoting overall well-being and reducing stress and anxiety among the animals.

In complex situations involving ethical dilemmas or medical decisions, the philosophy of care provides a reference point that helps staff make choices that are in the best interests of the animals. For example, if a situation arises where there is no clear protocol within the animal care manual, a philosophy of care can provide the guidance on how to make a critical decision that is in line with an organisation's mission and vision.

A philosophy of care is a fixed document, rather than a care manual, which is a living document. The care manual can respond to new situations, whereas the philosophy of care is written in general enough terms that it can inform each situation that arises.

How Can a Philosophy of Care Address the Pitfalls?

- *Addressing the pet mentality*

    A philosophy of care addresses the pitfall of a pet mentality by insisting on written care goals. A good philosophy of care would never have a care goal where staff gratification takes precedence over animal welfare. Rather, it would lay out what is important for the animal to receive in terms of exemplary care.

- *Addressing a lack of transparency*

    A philosophy of care addresses a lack of transparency that clearly defines why care protocols are enacted. This allows for staff, stakeholders, and the public to see why certain actions were carried out. It also lays out the framework for how mistakes are handled so that continuous improvement can occur.

- *Addressing issues with regulatory bodies*

    Similarly to how a philosophy of care addresses and lack of transparency, it can great help with feelings with regulatory bodies by clearly defining why a facility has taken the actions, or housed animals in a particular way by specifying the justification for every aspect of care. A well written philosophy of care can be referenced during interactions with regulatory bodies. This can lead to a more productive dialogue.

- *Addressing budgetary cuts that compromise welfare*

  A philosophy of care can help ensure that the necessary components of exemplary care, as defined by the philosophy, are never subject to austerity measures. For example, a philosophy may demand substrate, enrichment, or food of a certain quality. These, then, would be untouchable to budget cuts.

- *Addressing the propensity to enter into rivalries with other facilities*

  A philosophy of care can address unproductive rivalries with other facilities by ensuring that the focus of an organisation always remains on providing care and welfare to its residents, rather than an unhealthy competition. A well-written philosophy may even encourage collaboration rather than competition.

Implications

Animal care is challenging. Animal care is emotionally charged. Animal care is expensive. Animal care is controversial. While all of these things may seem obvious, they can all lead to pitfalls that can be an existential threat to any animal care organisation. Anyone engaged in a career in animal care will ultimately face many, if not all of these pitfalls at some point. However, the best defence against them is to (1) prepare for them, (2) recognise them, and (3) eradicate them.

To prepare for pitfalls, a highly effective measure is to create a strong philosophy of care that serves as a constitution of care for an animal care organisation. Once this philosophy is enacted, any pitfall that compromises the care dictated by the constitution should be recognised and dealt with.

Case Study: Ensuring the Standards of Care at Animal Sanctuaries

Kristin Leppert, Program Director, Global Federation of Animal Sanctuaries (GFAS)

The Global Federation of Animal Sanctuaries (GFAS) is the only worldwide recognized accreditation body for animal sanctuaries. GFAS's mission is to accredit and recognise sanctuaries and rescue centres, support them to achieve the highest Standards of Excellence, promote collaboration, and raise awareness of their work. Kristin Leppert joined GFAS early in 2019. Prior to GFAS, Kristin was the campaign director for Save Endangered Animals Oregon – a statewide coalition that successfully banned wildlife trafficking in the state in 2016. She worked in Washington, D.C. for several years where she led the Fur-Free Campaign for The Humane Society of the United States. Kristin directed investigations, influenced fashion design students and secured high-visibility sponsorships at New York Fashion Week. Her investigative research led prominent retailers Tommy Hilfiger, Calvin Klein and Kenneth Cole to adopt fur-free policies and supported the passage of the Truth in Fur Labelling Act in 2010. She has appeared in national media including The Today Show and Good Morning America. Kristin holds a degree in Political Science from Portland State University.

Figure 11.2 Kristin Leppert of GFAS. (Photo credit: Kristin Leppert's collection).

On Becoming Involved with GFAS

My heart has always been with animals, and when I knew that I wanted to work on their behalf, the natural place to start was at my local veterinary clinic. I still regard my days working hands on with the animals to be the very best of my life. When I became more aware of the industrial systems behind animal suffering, I started volunteering with The Humane Society of the United States on state legislative campaigns and then became an intern in the federal affairs office in DC. My internship led to positions leading campaigns on canned hunting and humane alternatives to fur in fashion.

Interest in animal sanctuaries began after helping an Iraqi activist gain attention for a starving lion and bear left to die in the abandoned Mosul Zoo. Seeing the animals' eventual rescue and first steps into the tranquillity of their new sanctuary home was transformative. It seems like the stars lined up soon after when a friend introduced me to the Global Federation of Animal Sanctuaries (GFAS).

On the Mission and Work of GFAS

The mission of GFAS is to accredit and recognise sanctuaries and rescue centres, support them to achieve the highest Standards of Excellence, promote collaboration, and raise awareness of their work.

My role as a wildlife program director is to work directly with the people that manage wild animal sanctuaries, and guide them in their pursuit of accreditation. This means that most days I'm reviewing animal care protocols and policies to do with housing, veterinary care, enrichment, nutrition, human safety, and more. Other days, I'm out on site visits on the other side of very secure barriers from elephants, tigers, chimpanzees, and all types of wild animals.

On Animal Sanctuaries

The definition of GFAS for a 'sanctuary' is any facility providing temporary or permanent safe haven to animals in need while meeting the principles of 'true sanctuaries' by providing excellent and humane care for their animals in a non-exploitative environment and having ethical policies in place. True sanctuaries hold non-profit status, and are committed to ethical policies such as no breeding, no commercial trade in animals or animal parts, animals are never exhibited anywhere or taken from the

Figure 11.3 Kristin Leppert examining the care and welfare of animals at a sanctuary. (Photo credit: Kristin Leppert's collection).

sanctuary or enclosures for non-medical reasons, and the public does not have direct contact with wildlife.

Unfortunately, there are commercial operators that describe themselves as a 'sanctuary' or 'rescue' and the quality of animal care and motive behind these organisations varies widely among them. We often refer to the worst of these places as 'pseudo sanctuaries' where the owners are out to make a profit from selling animal encounters like cub petting and photos with animals. Pseudo sanctuaries are notorious for breeding and buying animals at auctions, cramped and dismal living conditions, lack of proper veterinary care, and poor diets.

On How GFAS Oversees the Standards of Care at Animal Sanctuaries

We are the only globally recognised organisation for certifying that a facility meets the highest standards and recognises those as a 'true sanctuary.' When an organisation applies to become accredited, they must endorse our core principles, fill out an animal care application for all the species they care for, and submit to a very transparent process of providing policies, protocols, and financial and governance documents. Interviews are conducted and a site visit to the sanctuary is completed to assess compliance in areas of animal well-being, governance, housing, veterinary care, nutrition, enrichment, and safety.

Every accredited organisation must undergo a renewal process every three years that includes submitting updated information and a site visit. Anyone with concerns about a sanctuary can submit a complaint through the GFAS website and these are thoroughly investigated through interviews and additional site visits.

On the Biggest Challenges She Has Faced in Her Career

Animal protection is a challenging field to work in as there are wildly different views on what it looks like to truly protect animals. At every turn in my career working on behalf of animals, I have been pressed on my beliefs, and choices, while sometimes uncomfortable, it's fair. There's always going to be more to learn and more things to let go. Like all sectors in the animal movement, the animal sanctuary field is growing and evolving and it requires us to be open, inquisitive, and adaptable to change as we learn more about our own true nature and that of the animals we serve.

References

Chalupova, M., Voracek, J., Smrcka, F. and Kozakova, P., 2014, Dynamic modelling of zoo management: from challenge to opportunity. In *European conference on management* (pp. 24–33). Reading: Academic Conferences and Publishing International Limited.

Mills, K. E., Han, Z., Robbins, J. and Weary, D. M., 2018. Institutional transparency improves public perception of lab animal technicians and support for animal research. *PLoS One*, *13*(2), p. e0193262.

Conclusion

Curatio Fundamentorum 7: The care of living things is a science and must be approached scientifically.

12 The Science of Shipbuilding

Figure 12.1 Sea Lion at the Memphis Zoo. (Photo credit: author).

A shipbuilder, embarking on the act of creating a seafaring vessel that can be navigated through the ocean and travel from one destination to the other while being able to withstand stormy seas and unpredictable tides, must have a keen understanding of all aspects of boating. He must understand buoyancy. He must understand how ocean currents work. He must understand the effects of saltwater on wood. He must understand the properties of wind, and how wind can direct a sail. He must understand how a rudder works.

But this isn't all.

A shipbuilder must understand how a boat can house a crew. He must understand how a crew can navigate and operate the ship. He must understand how a kitchen can be constructed within a boat so that the crew can stay fed. He must understand that the crew is going to produce trash and waste, and understand how to dispose of it all in a safe way. He must understand that sometimes the crew might get sick and require a clinic. He must understand that there are times when emergencies happen, and the crew may need lifeboats. There is a science to shipbuilding, and as such, it must be approached scientifically.

A scientific approach means relying on what has worked to build ships in the past. One should consult shipbuilding manuals. One should ask other shipbuilders what worked for them. And one should experiment and document on their own how to build the best possible ship they can.

Imagine if a shipbuilder did not take a scientific approach to constructing a seafaring vessel. Imagine if a shipbuilder decided that all they require is a love of the sea and an appreciation of the beauty of boats. Imagine they built a ship based on gut feelings and aesthetics. Imagine being part of the crew on that ship. Obviously, whoever boarded that ship would face imminent danger and a destiny that exists somewhere at the bottom of the ocean floor.

And so it is with the care and welfare of animals. Anyone who is creating an environment where animals can thrive, designing protocols where animals can thrive, and carrying out practices that allow animals to thrive, is engaged in a scientific undertaking, and as such, must approach it scientifically.

Every individual directly or indirectly managing the care of an animal is a care technician, and must be well versed in the science of animal care and welfare. Every care technician, regardless of their specific role, is fundamentally responsible for the welfare of the animals they oversee. This responsibility is manifested through consistent care practices, informed by objective evidence and supported by rigorous training and experience. They must rely primarily on written protocols to ensure uniformity in care, which is crucial for maintaining the health and well-being of animal populations.

Each care technician must define what thriving means for each individual in their care. They must pursue this definition through a keen knowledge of the species, the needs of the species, and how that species responds to the confines of managed care. They must possess a knowledge of the individual and proceed from the idea that each individual is unique, with unique needs, challenges, preferences, anxieties, and histories.

Each care technician must help to create an environment where these individuals can thrive. They must understand that an environment is not just the physical landscape that an individual occupies. Rather, it is the sum of all external factors that can affect an individual in any way. This environment must meet the species' needs and individual needs of its residents – providing a species-specific and individually specific richness that allows living life to its fullest potential.

Each care technician must understand that there are internal needs and challenges for each animal. These internal needs can be psychological and emotional. They can also be physiological. As such, they must have the tools to diagnose challenges and the tools to treat what they diagnose. They must have ways of assessing and treating that limit the amount of stress on the individuals they are treating.

Each care technician plays a factor in the culture of those that share a role in the care of these animals. They must contribute to a culture of safety, a culture of respect for each other, and a culture of transparency. They must understand the emotional toll that taking care of animals can have on themselves and their colleagues, and know the resources to help themselves and each other to navigate these emotions. They must adhere to a philosophy of care that ensures that their culture is always focused on the welfare of the animals they manage.

Only through a scientific approach can a care technician be assured that proper animal welfare can be achieved. Just as a ship without a rudder is doomed to careen into an island or iceberg, a care technician without science is failing the animals in their care. However, a care technician that is armed with the science of animal care and welfare can successfully navigate the perilous and unpredictable waters of animal husbandry with a resilience that is based on evidence and knowledge.

The animals in our care are individuals who have, by reasons of birth or capture, been taken out of their natural environments by forces beyond their control. We, as care technicians, are given the ultimate responsibility to find means of thriving – living life to its fullest species-specific potential – in a world that we have created for them. They are, for lack of a better example, castaways in a strange world. Even those that are born in an artificial environment are foreigners to it by their very nature. It is our task to both create an environment where thriving is possible, and guide them, through our protocols, to all of the opportunities that the environment provides. Then, we must allow them to freely live in this environment without unnecessary interference from us.

A lighthouse keeper provides a beacon of light that safely guides a ship to safer tides. He guides it to a place where the ship can both survive and *freely* sail as it was built to do. As animal care technicians, we are like lighthouse keepers. We provide this beacon of light, through our daily routines, to the animals in our care. We guide them to a world that we have curated. In this world there are niches – physical, behavioural, and ecological spaces where they fit. It is a world that transforms castaways into building blocks of an expanding environment.

– a world curated just for them.

Case Study: A Career in Primate Behaviour

Dawn Abney, Manager, Laboratory Animal Medicine at Charles River Laboratories

On Her Career Journey
My journey into the career of primate behaviour might be a bit unique. It all started when I was 10 years-old and watched the movie 'Project X' with Matthew Broderick

Figure 12.2 Dawn Abney at Charles River Laboratories. (Photo credit: Dawn Abney's collection).

and Helen Hunt about the NASA space programme using chimpanzees. In the movie, a chimpanzee was taught sign language and I thought that was the most amazing thing I had ever seen. My love of primates started that moment. I remember immediately telling my mom I was going to grow up to work with apes and monkeys, and she was convinced I was going to work at a zoo or circus! The world of primate behaviour in a research setting never occurred to either of us.

Between eighth grade and my senior year of high school, I volunteered at the Dickerson Park Zoo (DPZ) in Springfield, MO, so I could start working with animals. The DPZ only had a handful of primates, so I mostly worked with Asian elephants and big cats, but I loved every minute of it. I was there almost every Saturday for five years and learnt so much about exotic animals. For college, my extremely supportive mother, did a lot of research to find out which universities in the U.S. had primate centres associated with them. During spring break of my junior year in high school, we went on a large tour of several universities and their primate centres. I am so blessed to have entered this field with a background knowledge and awareness of the many primate centres scattered across the country. I have referred to this experience many times throughout my career when talking to colleagues at those centres, since I visited them in high school! I finally landed at The University of Texas at Austin because, (1) I was accepted! and (2) The MD Anderson Primate Research Centre was located in Bastrop and had chimpanzees!

My experience at MD Anderson absolutely shaped the rest of my career. This was my first experience with primates in a research setting, and I fell in love. It is such a privilege to contribute to life-saving research and science while studying animal behaviour and welfare. After college, I moved to Washington, DC and briefly worked at Walter Reed Army Medical Centre as an animal care tech and then moved over to the National Institutes of Health – Division of Veterinary Resources as a Behavior Tech. While at NIH, I was able to work with a large variety of lab animal species, including a variety of primates. I had my first experience with New World Primates

Figure 12.3 Dawn receiving an award for her work at Charles River Laboratories. (Photo credit: Dawn Abney's collection).

(owl monkeys and squirrel monkeys) as well as baboons, African Greens, and three different species of macaques. I guess you could say, this is where I 'cut my teeth' as a young primate behaviourist. For two and a half years, I was the only Behavior Tech at the Poolesville, MD site and learnt the most from my animals! I worked on positive reinforcement training for blood collection, with monkeys who had serious abnormal behaviours and needed specific treatments, and learnt so much from the dedicated veterinary and husbandry care staff. I truly enjoyed my time at NIH.

Arriving at Charles River Laboratories

In 2007, I moved to Reno, NV, and became the Behavior Management Specialist at the Charles River Laboratories (CRL) facility. At that time, the site was both a CDC import quarantine site and conducted safety assessment studies as a Contract Research Organization (CRO). We had approximately 4,000 monkeys (cynomolgus and rhesus), but two Behavior techs! I was the third! My main priority at the Reno site was to develop a proactive behaviour and enrichment programme. I wanted to move away from only reacting and treating abnormal behaviours to preventing them. Using my experience from MD Anderson and NIH, I worked with the veterinary team to build a robust Behavior Management Program that would meet the species-typical needs of our amazing animals. The path to developing and building a multifaceted programme took a lot of support and people higher than me in the organisation supporting my suggestions and recommendations. Sometimes the 'yes' came easy, and sometimes it took years or a change in management to get the 'yes', but if it was important for the animals, it was important to me!

During my time at CRL, I have grown in my role and responsibilities. I started as the supervisor of two Behavior techs and was able to grow the Behavior Team to four techs (five including me!). We were responsible for social housing of all monkeys onsite, behavioural observations, ensuring animals had appropriate enrichment items, ordering/receiving of all food and toy items, and providing training to staff on NHP behaviour. After approximately three years, I was promoted to the role of Senior Supervisor of Behavior Services, which included overseeing the Vet Techs, Surgery Techs, and Technical Training Department. This expanded my ability to ensure great animal welfare and care by working closely with these various teams to promote animal behaviour training and positive human behaviour when working with primates. I continued in this role, as well as expanded it, by adding an area of group housed primates and continuing to grow the Behavior department with additional technicians.

In 2021, I was promoted to my current role as the Associate Director of Animal Health and Welfare. In this role, I continue to oversee many of the same subgroups within the Laboratory Animal Medicine department, but I also get the opportunity to work with many of the other CRL NHP sites to improve and enhance their behaviour and enrichment programmes. One of my biggest accomplishments of my career has been building a Behavior Medicine Program at the Reno site that is an example to the other CRL sites. Our programme is often praised by our clients, regulatory agencies, and others from visiting sites. Being able to push the boundaries of welfare and implement novel enrichment ideas or change techniques of how we handle and work with primates has been incredibly rewarding.

Obstacles and Challenges

But, I have certainly met my fair share of obstacles and challenges along the way. Pushback to new ideas or challenges to changes you want to implement, comes in many forms! When I first started at the site, our monkeys were only receiving eight hours of full social housing contact a week. This means they were spending the majority of their week with no access or grooming access. Primates are a highly social species, so eight hours of full social contact a week was unacceptable. The main reason for this was the inability to collect individual animal data (e.g., food consumption, observations of faeces or other clinical findings). I had to work with our Toxicology and Operations management to break down those fears and find processes where we could still identify if an individual animal was experiencing a clinical symptom while keeping the animals socially housed. Through the work of several individuals, we were able to successfully make the change to full social housing at all times and only to separate animals when absolutely necessary, and for brief periods of time.

Balancing Welfare and Science

My staff and I often joke that we push many boulders up the hill on a daily, weekly, monthly, and yearly basis! We take our job to be the advocate and voice of the animal very seriously, and if we are passionate about an animal welfare-related issue, we will continue to be persistent in making change happen. Working in a research environment, we have to balance welfare and science on a daily basis. My job is to ensure animal welfare always has a resounding voice at the table. Animal behavioural health is as important as animal clinical health, and they are closely linked. Advocating for the behavioural health of the animal through the Behavior Management Program, as well as through the other groups I oversee, is a big responsibility, but also a great source of pride and joy in my career. I feel blessed every day to get to work with the beautiful animals who ultimately give their lives so we can have life-saving medicines. My goal is to uphold my staff, my site, my company, and my industry to the highest welfare standards possible and always push the boundaries for making my animals' lives better!

Glossary

A priori Pre-observational facts.

Action-oriented system A process whereby a mental perception causes an action either by motivational urge or involuntary response.

Ad libitum Narrative descriptions.

Affect The emotional experience associated with the fulfilment or deprivation of a need.

Affective neurological processes Mental processes that create emotional states.

Affective neuroscience The field of research that focuses on studying the neural basis of emotions in both humans and animals.

Anecdotal observations Simple accounts of occurrences.

Anthropogenic change Environmental change that is caused by humans.

Anticipatory behaviour The act of perceiving something within, or just outside, one's environment and anticipating what comes next.

Artificial environment Any animal environment where any aspects of its anatomy have been replaced and replicated by human agency.

Basic needs The minimum resources one needs to survive.

Benchmarking A methodological approach used to assess animal welfare involving comparing the welfare of animals in a particular setting to a carefully chosen reference population.

Biological functioning perspective Focusing on changes in physiological responses or behavioural patterns.

Body condition scoring (BCS) A tool that enables a care technician (along with a veterinarian) to determine the relative health status of an animal by observing the overall state of an animal's body.

Chi-square A statistical test that measures the distance between our actual results and expected results.

Circadian rhythms Various physical, behavioural, and mental processes that a species undergoes during a 24-hour cycle; dictated by a species' ecological time niche.

Circular change Environmental change that is periodic, regular, and cyclical.

Cognition The mental process, the perception and awareness that is created, and the conditioning that comes from it. Cognition is a genetic trait that is enhanced through learning.

Cognitive stimulation What occurs each time a cognitive process is engaged? Cognitive stimulation affords the animal more skills to survive. Cognitive stimulation is integral for survival.

Conditioning Animals store a mental process and reaction to a given stimuli or scenario.

Communicative movements Movements that are intended for gestural or visual communication.

Compassion fatigue The physical and/or mental exhaustion, coupled with an emotional withdrawal that affects people in caregiving roles.

Culture Learnt behaviour that is passed down and becomes fixed within a population.

Culture of safety The collective values, attitudes, and practices that prioritise safety as a fundamental and continuous concern in every aspect of facility operations.

Cultural traits The various facets of a culture that make up the collective culture.

Data sheet The method of collection whereby the observer records ethological data in a structured manner.

Deep litter substrate A method of substrate whereby an enclosure is designed to hold several thick layers of substrate that break down wastes so effectively as to only require spot cleaning alongside a periodic complete replacement schedule.

Dynamic modelling A method of budgeting that takes into account how systems such as sources of revenue and financial needs change over time.

Ecological time niche The manner in which a species has evolved to behave at any given moment during a 24-hour cycle; determining a species circadian rhythms.

Enrichment programme Providing prescribed stimulation for animals in artificial environments.

Ethogram A list of event behaviours and state behaviours that one may reasonably observe within a given period of time relative to the species and circumstance.

Ethology The process of analysing how each behavioural instance creates an overall behavioural state of affairs.

Event behaviours Instantaneous behaviours that occur as a short momentary occurrence.

Extrinsic motivation An action that results in an immediate and clear ecological reward, such as eating, breeding, or fleeing danger.

Facultative needs Needs that can change within the course of an individual's lifetime.

Fight or flight response A physiological and behavioural response to danger.

The Five Freedoms An internationally recognised list of five aspects of animal welfare for animals in captivity.

Flow The balance between the appropriate level of skill that a species possesses to meet the specific challenges of an environment.

Focal sampling The observer surveys one individual and records which behaviour, listed in the ethogram, is being exhibited.

Functional equivalency The degree to which an artificial habitat replicates, or synthesises, a functional aspect of a natural habitat relative to the particular species.

Goodness of fit test Statistical tests that measure the distribution of actual results versus expected results.

Hallmark traits Evolutionary traits that define a species.

Hibernation An extreme reduction in physiological activity that can occur for weeks or even months.

Holistic A multidisciplinary and multifaceted approach to accomplishing a task. Animal care is a holistic science that includes behavioural, veterinary, husbandry, and physical design approaches.

Intake sheet A daily record of all food items an animal consumes along with the calorie count of each item.

Intrinsic motivation A motivation to perform an action for the sake of doing it with no clear or immediate reward.

Intrinsic reward The satisfaction of performing an intrinsic action.

Invasive techniques Assessment methods that require penetrating the body of an individual.

***K*-selected species** Species that live in relatively stable environments and, as a result, they have single-offspring births.

Life history traits The natural and species-specific stages of an individual's lifetime.

Linear change An environmental shift that persists.

Locomotor movements Travelling movements.

Maintaining behaviour Routinely practising a conditioned behaviour that has already been shaped.

Manipulative movements Movements intended to manipulate an object or part of an environment for any given reason.

Microclimates Small areas of a perpetual climate variance within a larger habitat.

Natural ecology The interactions with the natural environment in which an individual lives, both internally and externally.

Needs What one requires to both sustain and utilise one's physical and behavioural adaptations.

Negative acclimation Acclimation that creates chronic welfare issues.

Neurological process A process, by which an external factor is perceived by an individual, initiates an action-oriented response, and determines a behaviour and emotional state.

Non-invasive techniques Assessment methods that are largely observational or can be performed without delving into the body of an individual.

Non-locomotor movements Movements that are intended for any other reason besides locomotion.

Operant conditioning Modifying behaviour by reinforcement.

Philosophy of care A foundational framework that outlines the core principles, values, and approaches an animal care facility adopts to ensure the well-being and ethical treatment of the animals under its care.

Positive reinforcement training A process involving shaping and maintaining animal behaviours by offering an immediate reward for desired behaviours.

Punishment The process of adding negative stimuli, or removing a positive stimulus, if the undesired behaviour is exhibited.

Quality of life assessment An empirical tool to measure observable signs of well-being that collectively point to an individual's quality of life.

***R*-selected species** Species that live in unstable environments and, as a result they have multiple offspring births (giving them a greater chance of reproductive success).

Range of mental processes The array of mechanisms by which a species has evolved to cognitively process the environment around them.

Range of movement The types of motion within a given natural habitat that a species has evolved to exercise. This includes ways an animal utilises extremities, methods of locomotion, and geographic spatial coverage within the lifetime of an individual.

Reinforcers Conditioning stimuli that influence your actions.

Resiliency The ability of an animal to adapt to any given change in an environment.

Rich social environment An environment that most closely reflects the social makeup of a natural population of a species

Satiance The feeling of well-being due to an absence of hunger.

Scan sampling, The observer surveys a group of animals, and records which behaviour, listed in the ethogram, is being exhibited.

Self-maintenance movements Movements that are intended to satiate a personal need.

Sexually philopatric One sex disperses from a social group while the other sex remains within the group.

Shaping behaviour Conditioning the steps necessary in order to reach the larger behavioural goal.

Socially interactive movements (non-communicative) Movements means for any social interaction with conspecifics, which would include fighting, grooming, parenting, and breeding.

Speciation A new species arises out of a parent population.

State behaviours Behaviours that occur for a measurable amount of time.

Subspeciation A new subspecies arises out of a parent population.

Substrate Substance that absorbs, breaks down, and controls the spread of waste, until such time as an enclosure can be cleaned.

Thriving The constant spacious pursuit of living life to its fullest potential. In animal care, we look to ensure that all individuals are given all available avenues to a life of thriving.

Torpor A temporary reduction in physiological activity.

Training bridge A signal that is used to communicate to the animal that the specific behaviour is correct and will be rewarded.

Transient Individuals that frequently move between social groups.

Index

For EU product safety concerns, contact us at Calle de José Abascal, 56–1°,
28003 Madrid, Spain or eugpsr@cambridge.org.